THE
HUNGRY
BRAIN

decide whether that pastry behind the counter looks and smells tempting. System 2 processes are slow, effortful, rational, and conscious. They decide whether the potential health and weight consequences of the pastry are worth it, and whether or not to override the impulse to buy it. As this example illustrates, these two systems can harbor competing goals within the same brain. But system 1, Kahneman argues, is the more influential of the two in our everyday lives.

Kahneman's work is part of a growing body of psychology and neuroscience research that's gradually chipping away at the domain of the conscious, rational brain. What's emerging is a picture of our decision-making process that's far less conscious, and more impulsive, than most of us intuitively believe.

Very few of us *want* to overeat. And certainly none of us want to consistently overeat for ten, twenty, or thirty years, develop obesity, and end up with a high risk of diabetes and cardiovascular disease. The $60 billion diet and weight-loss industry is a powerful testament to the fact that most of us would rather not overeat. Yet judging by the fact that one-third of American adults have obesity, and another third is overweight, most of us do precisely that.

The fact that we want to eat less or eat healthier food, yet we often don't, is consistent with Kahneman's idea. It implies the existence of a conscious, rational brain that cares about abstract concepts like health, body weight, appearance, and their future trajectories. And it also implies the existence of a nonconscious (or minimally conscious), intuitive brain that cares about concrete, immediate things like that piece of double chocolate cake in front of you, and doesn't always listen to the sage advice of its rational "big brother."

The conflict between the conscious and nonconscious brain explains why we overeat even though we don't want to. Although we try to control our behavior using the conscious parts of our brains, the nonconscious parts work to undermine our good intentions. Returning to the *Dietary Guidelines for Americans,* the problem isn't that the information is wrong—it simply targets the wrong brain circuits. If that's true, then what circuits are actually in charge of our everyday eating behavior, and how do they work? If we can answer those questions, we can understand why we make destructive food choices in the modern world, and how to stop.

the advice of the *Guidelines,* our intake of soda, refined sugar, refined starch, added fats, and highly processed food skyrocketed, along with our calorie intake.

I doubt these negative changes had much to do with the *Guidelines;* rather, they were the result of gradual but profound socioeconomic forces that altered the way we interact with food.[3] Yet they do point to a fundamental aspect of human nature: Information alone isn't always an effective way to change behavior. I believe the failure of the *Guidelines* to prevent the obesity epidemic reflects a broader misunderstanding of human eating behavior—one that undermines our ability to manage our weight effectively, both as a nation and as individuals.

It's quite rational to care about your health, and therefore how you eat, because health has a major impact on your well-being, life span, and performance in many areas of life. If our eating behavior is primarily guided by conscious choices based on rational thinking, then educating the nation on what to eat should be a highly effective way of making us slimmer and healthier over time. The *Guidelines* assume, as many of us do, that if we just have the right knowledge about how to eat the right foods in the right amounts, we'll do it. In contrast, if our everyday eating behavior is primarily guided by brain systems that aren't so rational, information alone shouldn't be a very effective way to change it, no matter how accurate, clear, and compelling it is.[4] I think our nationwide experiment over the last thirty-five years supports the latter scenario.

To understand why, we must turn to the brain. The human brain evolved over more than five hundred million years as an information-processing organ that supports our survival needs. It's an incredibly complex piece of biological machinery, but we can conceptualize some of its functions in simple ways. In his fascinating book *Thinking, Fast and Slow,* Nobel Prize–winning psychology researcher Daniel Kahneman divides the brain's thought processes into two domains he calls *system 1* and *system 2*. System 1 processes are fast, effortless, intuitive, and nonconscious. They

---

3 For example, increasing affluence, the gradual shift to two working parents rather than one, and the expanding influence of the processed food and restaurant industries. For more information on these topics, see the books *Salt Sugar Fat, Fast Food Nation, Food Politics,* and *The World Is Fat.*

4 Although this depends on the individual; it does have a large impact on behavior in a small segment of the population.

eat minus the number you burn, and it lays out its four-point weight management strategy:

- Increase physical activity
- Eat less fat and fatty foods
- Eat less sugar and sweets
- Avoid too much alcohol

Seems pretty reasonable. Yet you know what happened next: We got fatter. Between 1980 and today, the US obesity rate more than doubled. Certain diet contrarians have seized on this correlation, suggesting that the *Guidelines* themselves made us fat by recommending that we replace dietary fat with carbohydrate, leading us to eat too much refined starch and sugar. Yet the evidence has never supported such speculation, in part because Americans who actually follow these recommendations tend to be leaner than Americans who don't.[1] Not to mention the fact that the *Guidelines* specifically emphasize limiting sugar and other refined foods.

The *Guidelines* may not have been flawless, but that's not why they failed to constrain our expanding waistlines. The reason they failed is that we simply didn't listen. We didn't listen in 1980, and we have largely ignored the advice from the updated versions that have appeared every five years, despite the fact that they are now widely taught in public schools. It's not that we didn't change at all. We did start drinking less whole milk, replacing it with low-fat and skim milk (for better or worse),[2] and we replaced some of our beef intake with chicken, which is lower in saturated fat. But we were extremely selective about what we implemented. Against

---

1 Two additional lines of evidence undermine this idea. First, our absolute intake of fat didn't actually decline, according to USDA and CDC data. It only declined on a percentage basis because we began eating more carbohydrate. Depending on which data set you cite, our absolute fat intake either increased (USDA) or remained approximately the same (CDC). Second, the USDA has been publishing dietary guidelines in one form or another since 1894, and the 1980 version wasn't the first one to recommend limiting fat intake.

2 My colleagues Mario Kratz and Ton Baars and I published a paper in 2013 that reviewed studies on the association between full-fat dairy intake and obesity, metabolic health, and cardiovascular disease. As far as I know, we were the first to do so. As it turns out, people who drink whole milk tend to be leaner and metabolically healthier than those who don't, and the two groups also have about the same cardiovascular risk. I think it's too early to know exactly what the findings mean, but they certainly raise questions about why nutrition professionals recommend reduced-fat dairy.

Introduction

n 1980, the US Department of Health and Human Services and the US Department of Agriculture released a document titled *Dietary Guidelines for Americans*. The *Guidelines* were intended to reduce the risk of obesity and chronic disease by offering simple, evidence-based dietary recommendations for American policy makers, health professionals, and the general public. The document was only twenty pages long and focused on the following seven goals:

1. Eat a variety of foods
2. Maintain ideal weight
3. Avoid too much fat, saturated fat, and cholesterol
4. Eat foods with adequate starch and fiber
5. Avoid too much sugar
6. Avoid too much sodium
7. If you drink alcohol, do so in moderation

If you followed the *Guidelines* faithfully, you would end up eating a diet that's not very different from what most nutrition professionals recommend today: primarily whole grains, beans, potatoes, vegetables, fruits, nuts, lean meat, seafood, and dairy, with little added fat, sugar, or highly processed foods.

In the section titled "Maintain Ideal Weight," the *Guidelines* explain that changes in body fatness are determined by the number of calories you

THE
HUNGRY
BRAIN

Contents

Introduction   1

1 The Fattest Man on the Island   7

2 The Selection Problem   23

3 The Chemistry of Seduction   41

4 The United States of Food Reward   69

5 The Economics of Eating   87

6 The Satiety Factor   113

7 The Hunger Neuron   145

8 Rhythms   177

9 Life in the Fast Lane   197

10 The Human Computer   215

11 Outsmarting the Hungry Brain   223

Acknowledgments   239

Notes   241

Index   279

THE HUNGRY BRAIN. Copyright © April 2017 by Stephan Guyenet. All rights reserved. Printed in the United States of America. For information, address Flatiron Books, 175 Fifth Avenue, New York, N.Y. 10010.

www.flatironbooks.com

The Library of Congress Cataloging-in-Publication Data is available upon request.

ISBN 978-1-250-08119-3 (hardcover)
ISBN 978-1-250-08123-0 (e-book)

Our books may be purchased in bulk for promotional, educational, or business use. Please contact your local bookseller or the Macmillan Corporate and Premium Sales Department at 1-800-221-7945, extension 5442, or by e-mail at MacmillanSpecialMarkets @macmillan.com.

First Edition: February 2017

10   9   8   7   6   5   4   3   2   1

THE
HUNGRY
BRAIN

Outsmarting the Instincts
That Make Us Overeat

STEPHAN J. GUYENET, Ph.D.

FLATIRON
BOOKS
NEW YORK

Why do we have brain functions that seem to exist only to make us fat and sick? In fact, these functions evolved to help us survive, thrive, and reproduce—in an ancient world that no longer exists for us. We'll explore the lives of modern-day hunter-gatherers living similarly to how our ancestors would have lived for the last several million years, and we'll see how the same impulses that get us into trouble today were extremely useful to our forebears. While a calorie-seeking brain is an asset when calories are hard to come by, it's a liability when we're drowning in food.

Scientists call this an *evolutionary mismatch;* in other words, a situation in which once-useful traits become harmful once they're dragged into an unfamiliar environment. Many researchers believe that evolutionary mismatches explain a number of the chronic disorders that are so common in the affluent world. In this book, I'll argue that overeating and obesity are caused by a mismatch between ancient survival circuits in the brain and an environment that sends these circuits the wrong messages.

I'll be your guide on this journey through the science of overeating. I've always been fascinated by the brain, because it makes us who we are—and it also happens to be the most complex object in the known universe. After completing a BS in biochemistry at the University of Virginia and a Ph.D. in neuroscience at the University of Washington, I became interested in the role of the brain in obesity. I was driven by the perplexing question: Why do we gain excess fat, even though it's clearly not good for us? I joined the lab of Mike Schwartz, also at the University of Washington, as a post-doctoral fellow, and together we worked to unravel the mysterious neuroscience of body fatness. It quickly became obvious to me that we were studying the right organ: The brain is in charge of appetite, eating behavior, physical activity, and body fatness, and therefore the only way to truly understand overeating and obesity is to understand the brain. Researchers know quite a lot about how these processes work, but this knowledge is conspicuously absent from popular theories of obesity because most of it remains sequestered in the dusty pages of academic journals.[5] I aim to rectify that.

The nonconscious brain that guides our everyday eating behavior isn't just one system; it's a collection of many unique systems that reside in distinct areas of the brain. Modern research has given us incredible insights

---

5 Only dusty metaphorically, because it's mostly electronic now.

into these systems and how they affect our eating behavior, yet much of this information is only accessible to people with extensive science backgrounds. In this book, I'll decode these findings and take you on a tour of some of the most influential brain systems that drive us to overeat. Along the way, you'll learn about how your brain works in general, and you'll hear the voices of many remarkable researchers. I also translate this information into simple strategies you can use to manage your waistline constructively and painlessly.

Welcome aboard; I hope you enjoy *The Hungry Brain*!

1

THE FATTEST MAN ON THE ISLAND

S tout but not quite obese, and boasting a prominent belly, Yutala would have been an unremarkable-looking man in many places.[6] He would not have stood out on the streets of New York, Paris, or Nairobi. Yet on his native island of Kitava, off the coast of New Guinea, Yutala was quite unusual. He was the fattest man on the island.[7]

In 1990, researcher Staffan Lindeberg traveled to the far-flung island to study the diet and health of a culture scarcely touched by industrialization. Rather than buying food in grocery stores or restaurants like we do, Kitavans used little more than digging sticks to tend productive gardens of yams, sweet potatoes, taro, and cassava. Seafood, coconuts, fruits, and leafy vegetables completed their diet. They moved their bodies daily and rose with the sun. And they did not suffer from detectable levels of obesity, diabetes, heart attacks, or stroke—even in old age.

As extraordinary as this may sound to a person living in a modern society beset by obesity and chronic disease, it's actually typical of nonindustrialized societies living similarly to how our distant ancestors might have lived. These societies have their own health problems, such as infectious

---

6 Name changed for privacy.

7 Highest body mass index of all Kitavans examined by Lindeberg.

disease and accidents, but they appear remarkably resistant to the disorders that kill us and sap our vigor in affluent nations.

As it turns out, Yutala wasn't living on Kitava at the time of Lindeberg's study; he was only visiting. He had left the island fifteen years earlier to become a businessman in Alotau, a small city on the eastern tip of Papua New Guinea. When Lindeberg examined him, Yutala was nearly fifty pounds heavier than the average Kitavan man of his height and twelve pounds heavier than the next heaviest man.[8] He was also extraordinary in another respect: He had the highest blood pressure of any Kitavan examined by Lindeberg. Living in a modern environment had caused Yutala to develop a modern body.

Yutala is a harbinger of the health impacts of industrialization.[9] His departure from a traditional diet and lifestyle, and subsequent weight gain, form a scenario that has played out in countless cultures around the globe—including our own culture, our own families, and our own friends. In the United States, we have a tremendous amount of information about the diet, lifestyle, and weight changes that accompanied this cultural transition. This will provide us with valuable clues as we piece together the reasons why our brains drive us to overeat, despite our best intentions. Let's start by examining how our weight has changed over the last century.

THE COST OF PROGRESS

In New Guinea, as in many other places around the globe, industrialization has triggered an explosion of obesity and chronic disease. If we look back far enough, we can see traces of the same process happening in the United States.

In 1890, the United States was a fundamentally different place from what it is today. Farmers made up 43 percent of the workforce, and more than 70 percent of jobs involved manual labor. Refrigerators, supermarkets, gas and electric stoves, washing machines, escalators, and televisions didn't exist, and motor vehicle ownership was reserved for engineers and wealthy

---

8 Calculated from Yutala's body mass index of 28 and the average male Kitavan body mass index of 20.

9 To be clear, some of these health impacts are positive, such as access to vaccines and antibiotics.

eccentrics. Obtaining and preparing food demanded effort, and life itself was exercise.

How common was obesity among our American forebears? To find out, researchers Lorens Helmchen and Max Henderson pored through the medical records of more than twelve thousand middle-aged white Civil War veterans and used their height and weight measurements to calculate a figure called the body mass index (BMI). BMI is basically a measure of weight that is corrected for height so we can compare weights between people of different statures. It's a simple measure that's commonly used to classify people as lean, overweight, or obese (a BMI below 25 is classified as lean; 25 to 29.9 is overweight, and 30 and above is obese). When Helmchen and Henderson crunched the numbers, they found something truly remarkable: Prior to the turn of the twentieth century, fewer than one out of seventeen middle-aged white men was obese.

The researchers then calculated the prevalence of obesity in the same demographic between 1999 and 2000 using data from the US Centers for Disease Control and Prevention. They found that it started at 24 percent in early middle age and increased sharply to 41 percent by retirement age.[10] Side-by-side comparison of the data from 1890 to 1900 and 1999 to 2000 yields a striking contrast (see figure 1).

This suggests that obesity was much less common in the United States before the turn of the twentieth century, just as it remains uncommon in traditionally living societies today. Although obesity has existed among the wealthy for thousands of years—as demonstrated by the portly 3,500-year-old mummy of the Egyptian queen Hatshepsut—in all of human history, it has probably never been as common as it is today.

Let's take a closer look at the last half century, because that's the period over which our data are the most reliable—and during which these numbers have changed most dramatically. In 1960, one out of seven US adults had obesity. By 2010, that number had increased to one out of three (see figure 2). The prevalence of extreme obesity increased even more remarkably over that time period, from one out of 111 to one out of 17. Ominously, the prevalence of obesity in children also increased nearly fivefold. Most of these changes occurred after 1978 and happened with dizzying speed.

---

10 Ages forty through forty-nine and sixty through sixty-nine.

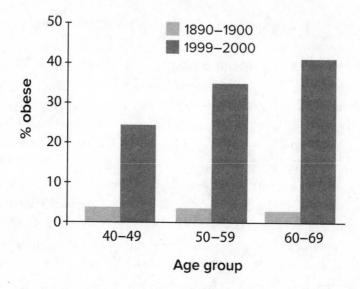

Figure 1. Obesity prevalence in white US men, 1890–1900 and 1999–2000. Data from
Helmchen et al., *Annals of Human Biology* 31 (2004): 174.

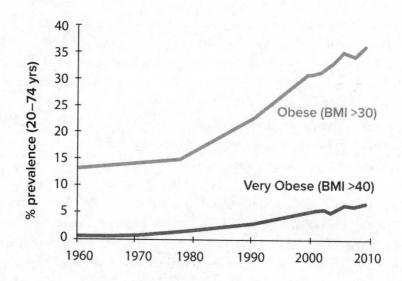

Figure 2. Obesity prevalence among US adults twenty to seventy-four years old,
1960–2010. Age adjusted. Data from Centers for Disease Control and Prevention NHES
and NHANES surveys.

Public health authorities call this the "obesity epidemic," and it's having a profound impact on health and well-being in the United States and throughout the affluent world. The latest research suggests that we may be gravely underestimating the health impacts of obesity, as up to one-third of all deaths among older US adults is linked to excess weight. Diabetes rates are soaring, as are orthopedic problems caused by obesity. Nearly two hundred thousand Americans per year are having their digestive tracts surgically restricted or rerouted to lose weight. Clothing is now available in staggering sizes such as XXXXXXXXL.

Why are we so much fatter than we used to be? The answer lies in what we've been eating and how it relates to the fat we carry, which we'll explore shortly. But we first have to understand how food delivers energy to our bodies.

THE CALORIE IS BORN

Contrary to popular belief, the term *calorie* was not invented by Snack-Well's. Rather, it was coined in the early 1800s and was used by scientists to measure energy in all its different forms by the same metric: as heat, light, motion, or the *potential energy* contained in chemical bonds. These chemical bonds are found in bread, meat, beer, and most other foods, which release their potential energy as heat and light when burned, just like wood or gasoline.

In 1887, Wilbur Atwater, the father of modern nutrition science, described how the potential energy in food fuels the furnace of the human body:

> The same energy from the sun is stored in the protein and fats and
> carbohydrates of food, and the physiologists to-day are telling us how it
> is transmuted into the heat that warms our bodies and into strength for
> our work and thought.

Recognizing the power of energy as a way to understand our bodies, Atwater's team was the first to exhaustively measure the calorie content of different foods by burning them in his energy-measuring "calorimeters."

When you see a calorie value on the side of your box of cereal, it was calculated using formulas Atwater developed by measuring the calorie content of food and adjusting it for the intricacies of human digestion and metabolism.[11] (The values are actually in *kilocalories*, or thousands of calories, which is denoted by capitalizing the word *Calorie*, a convention begun by Atwater.)

Atwater and his colleagues also constructed a giant live-in calorimeter to measure the combustion of food by the human body. This calorimeter was large enough to provide a modest living space for experiments lasting multiple days. Atwater's system was so effective that it was able to demonstrate with greater than 99 percent accuracy that the energy entering a weight-stable person as food is equal to the energy leaving the body. In other words, in a person neither gaining nor losing weight, the number of calories consumed is equal to the number burned.[12]

This statement can be rearranged as the *energy balance equation*:

Change in body energy = energy in − energy out

Energy enters the body as food, and it leaves as heat after we've used it to do metabolic housekeeping, pump blood and breathe, digest food, and move our bodies. We also use it to build lean tissues, such as muscle and bone, during growth. Any energy that's left over after the body has used what it needs is stored as body fat, technically called *adipose tissue*. Adipose tissue is the major energy storage site of the body, and it has an almost unlimited capacity. When you eat more calories than you burn, the excess calories are primarily shunted into your adipose tissue. Your *adiposity*, or body fatness, increases. It really is as simple as that, although as we'll see in later chapters, the implications are not as straightforward as they initially appear.

Atwater also discovered that chemical energy from different types of foods, including those rich in carbohydrate, fat, protein, and alcohol, is effectively interchangeable in the body: Roughly speaking, all calories are

---

11 The energy value of fat, digestible carbohydrate, and protein are approximately nine, four, and four kilocalories per gram, respectively.

12 Plus, of course, the few calories that end up in feces and urine; Atwater's finding really just means that the human body isn't a magical place that defies the laws of physics that govern everything else (in particular, it adheres to the first law of thermodynamics).

the same as far as the human furnace is concerned. More recent research has also supported the idea that the fat, carbohydrate, and protein content of foods has little influence on adiposity beyond the calories they supply. We know this because when researchers strictly control total calorie intake, varying the fat, carbohydrate, and protein content of the diet has no appreciable impact on adiposity—whether in the context of weight loss, weight maintenance, or weight gain. This undermines the commonly held belief that certain nutrients, like carbohydrate or fat, are more fattening than what their calorie content would suggest. Some foods are nevertheless more fattening than others, but this appears to be primarily because they coax us to eat more calories, not because they have a special effect on our metabolic rate.[13]

With this in mind, we can adapt the energy balance equation to describe long-term changes in adiposity:

Change in adiposity = food calories in − calories out

To gain fat, you must eat more calories, burn fewer calories, or both. To lose fat, you must eat fewer calories, burn more calories, or both. It's a simple concept, although applying it to weight loss can be surprisingly difficult, as many people know all too well.

If this principle is true, then we should expect to see that Americans began eating more calories, and/or burning fewer calories, as our waistlines expanded. Let's have a look.

CALORIES IN: HOW OUR CALORIE INTAKE HAS CHANGED

Measuring calorie intake in an entire country is a challenging task. Yet researchers have so far managed to do it in three different ways. The first

---

13 There is emerging evidence that certain diets, such as very low-carbohydrate, very low-fat, and high-protein diets, can modestly increase metabolic rate, but no study to date has shown that these effects result in meaningful differences in adiposity. It remains possible that diets at the extremes of fat and carbohydrate intake increase metabolic rate and accelerate fat loss slightly. Yet everything between these extremes, including typical low-carbohydrate and low-fat diets, seems to be approximately equivalent when calories are strictly controlled.

way is to measure food *production,* adjust for the amount that's being exported and imported, try to account for loss due to food waste, and see how many calories are left per person. The second way is simply to ask a representative sample of people what they eat and tally up the calories. The third way is to mathematically model the relationship between body weight and calorie intake, and use that model to calculate the change in calorie intake that should be required to produce the observed increase in weight.

I've graphed calorie intake estimates from all three methods in figure 3. As you can see, the methods yield different estimates, but they all agree that our calorie intake has increased substantially over the period of time that we gained weight the most rapidly (218–367 additional Calories per day between 1978 and 2006). The third method, pioneered by National Institutes of Health researcher Kevin Hall and highlighted in black on the graph, probably comes the closest to capturing the true increase in daily calorie intake over the course of the obesity epidemic: 218 Calories. Remarkably, this increase is single-handedly sufficient to explain the obesity epidemic that developed over the same period of time, without having to invoke changes in physical activity or anything else.

Right about now, you might be putting on your skeptic hat. *One pound of body fat contains about 3,500 Calories, so if we're really overeating by 218*

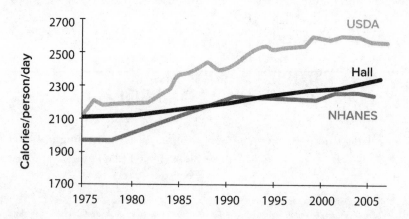

Figure 3. Calorie intake among US adults, 1975–2006. Data are from USDA Economic Research Service loss-adjusted food availability estimates, Centers for Disease Control and Prevention NHANES surveys, and Hall et al., *PLoS ONE* 4 (2009): e7940. Special thanks to Kevin Hall for providing raw data.

Calories per day, shouldn't we be gaining a pound of fat every sixteen days—twenty-three pounds per year—and requiring a forklift to get around after a decade or two? Actually, despite the popularity of this type of back-of-the-envelope math among popular media sources, public health authorities, doctors, and even some researchers, that's not how adiposity works. Hall and his colleagues have shown that this way of estimating changes in adiposity is way off target—and the consequences of this error have important implications for how we think about weight gain and weight loss.

The primary problem with this way of thinking is that it doesn't acknowledge the fact that as your body size changes, your body's energy needs change too. To illustrate the principle, think of your adipose tissue as a bank account. If you start out with $10,000 in savings, an income of $1,000 a month, and expenditures of $1,000 a month, in a year, your account will still contain $10,000. Now, imagine you get a raise and your earnings climb to $2,000 a month. At first, your lifestyle remains the same and you only spend $1,000 a month, saving the extra $1,000 per month. But gradually, you start to think it would be nice to have that new computer or fancy pair of shoes. You move into a nicer apartment. Your lifestyle expands, and your expenditures creep up. Six months after your raise, you're spending $1,500 per month, and after a year, you're spending the full $2,000 a month. Over this year, your bank account has been accumulating money, but at a gradually slowing rate, until accumulation stops when your expenditures match your income. Your account balance plateaus at about $16,000, and it remains there until your income or expenditures change.

And so it is for adiposity. When your calorie intake increases, your body weight increases, and this extra tissue burns calories.[14] Gradually, as your body enlarges, your calorie expenditure comes to match your extra calorie intake, and you reach a weight plateau. You're no longer eating more calories than you're burning, so your weight and adiposity stabilize at a higher level. The same plateau effect happens in reverse when a person cuts her calorie intake.

What are the practical implications of this? An important one is that it takes a larger change in calorie intake to gain—or lose—weight than most

---

14 The increase in metabolic rate is primarily due to the increase in lean mass (everything except fat) that occurs during weight gain.

people realize. Making small changes to your diet, such as cutting out one slice of toast per day, will lead to correspondingly small changes in adiposity that don't continue to accrue indefinitely. The new, evidence-based rule of thumb is that you must eat ten fewer Calories per day for every pound you want to lose. Yet it takes several years to arrive at a new stable weight, so most people will want to start with a larger calorie deficit to reach their target weight more quickly and then use the ten-Calorie rule of thumb to maintain the loss.

This offers a partial explanation for that scourge of conscientious dieters everywhere: the dreaded weight-loss plateau. This is where a person diligently cuts her calorie intake and successfully loses weight, but her weight loss stalls before she reaches her goal, even though she continues to follow her formerly successful diet. This phenomenon is real, and Hall's research offers two explanations for it. First, as a person loses weight, her smaller body needs less fuel, the calorie deficit gradually closes, and weight loss stalls. And second, weight loss ramps up her appetite, making it harder to maintain the calorie deficit (I'll explain why this happens in later chapters). To restart weight loss during a plateau, she must reestablish a calorie deficit, although that's easier said than done.

Three independent methods suggest that our calorie intake increased substantially over the course of the obesity epidemic, and this increase is sufficient to account for the weight we gained. Simply stated, we gained weight because we ate more.

Now, let's step back for a moment. We've been focusing on recent history because that's the time period over which we have the most complete information about adiposity. But what about the first half of the twentieth century? In figure 4, I've graphed our calorie intake over the last century, based on food disappearance data from the US Department of Agriculture (the second method of estimating calorie intake). These data are crude, but they serve to illustrate broad trends over time.[15]

As you can see, we ate more calories in 1909 than we did in 1960. Yet as far as we know, there was no obesity epidemic in 1909. Why not?

---

15 They're crude in part because food waste has increased over time, and they don't take that into account. This artificially inflates the recent increase in calorie intake (both in this graph and the previous one). That explains why the USDA's estimate of the increase in calorie intake is so much larger than the other two methods. Kevin Hall published a fantastic paper on this.

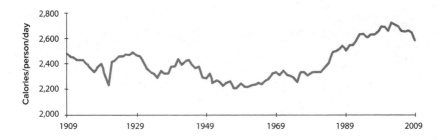

Figure 4. US calorie intake per person, per day, 1909–2009. Based on USDA Economic Research Service food availability estimates. Figure was derived from "nutrient availability" data by adding together calories from protein, carbohydrate, and fat. Data were adjusted for waste at a flat rate of 28.8 percent, and an artifact in 1999–2000 that resulted from a change in the assessment method for liquid oils.

CALORIES OUT: HOW OUR PHYSICAL ACTIVITY HAS CHANGED

This brings us to the second determinant of adiposity: the amount of energy leaving the body. Aside from metabolic housekeeping, pumping blood and air, and digesting food, another important thing we do with energy is to contract our muscles so we can walk, weed our crops, bale our hay, milk our cows, knead our dough, wash our laundry by hand, and put things together in a factory. As it turns out, we used to do those things a lot more a century ago than we do today. In other words, we ate a lot because we needed the energy to fuel our relatively high level of physical activity.

This is a critical point. In 1909, our high calorie intake was appropriate for our high calorie needs. As our lifestyles gradually became more mechanized over the course of the next half century, physical activity declined substantially. Fewer of us were working the fields with plows and hoes, and more of us were sitting behind a steering wheel, as demonstrated by the massive increase in automobile registrations shown in figure 5. Until 1913, fewer than one out of one hundred Americans had an automobile. Today, there are eight automobiles for every ten of us.

As we became a more sedentary nation, calorie intake declined appropriately until about 1960. People were working up less of an appetite, so they ate less.

Then, around 1978, something changed: We started eating a lot more

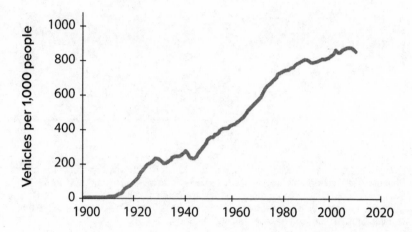

Figure 5. US automobile registrations per one thousand people, 1900–2010. Data from US Department of Energy.

calories—and our calorie intake continued to increase over the next twenty years, reaching historically unprecedented levels. Yet we remained sedentary. The obesity rate rose so quickly that public health authorities didn't realize what was happening until an epidemic was upon us.

A NATION OUT OF BALANCE

When calorie expenditure decreases and calorie intake increases, the energy balance equation leaves only one possible outcome: fat gain. We gained fat as we ate more calories than we needed to remain lean, given our physical activity level. In other words, we overate.

For most of human history, including the majority of the twentieth century in the United States, nearly everyone was able to approximately match their calorie intake to their calorie needs without even thinking about it. Yet mysteriously, since then, something has caused our calorie intake to uncouple from our true needs. Something pushed us to overeat.

What drove us to overeat? If we can answer that question, we'll empower ourselves to do something about it. Let's begin to answer it by asking another question: What's the most effective way to cause overeating?

ONE WEIRD TRICK TO MAKE A RAT OVEREAT

Obesity research has a long, rich history of fattening rodents for science. In the 1970s, researchers were looking for a better way to fatten their rats so they could study the development and impacts of obesity more efficiently. In the early days, researchers would simply add fat to standard rodent chow. This worked, but it often took months to produce a portly rat—making rodent obesity research a costly and time-consuming endeavor.

One fateful day when Anthony Sclafani, now the director of the Feeding Behavior and Nutrition Laboratory at Brooklyn College, was a graduate student, he happened to place a rat onto a lab bench where a fellow student had left a bowl of Froot Loops cereal. The rat waddled over and began to eat it heartily. This was surprising because rats are typically cautious with unfamiliar foods. Watching the rat greedily devouring human food, it occurred to Sclafani that foods marketed for people might be more fattening than the high-fat rodent chow he was currently using.

To see if he could design a faster and more effective way to fatten rats, Sclafani went to the supermarket and bought a variety of calorie-dense "palatable supermarket foods," including Froot Loops, sweetened condensed milk, chocolate chip cookies, salami, cheese, bananas, marshmallows, milk chocolate, and peanut butter. When Sclafani placed these foods into the rats' cages, along with the obligatory standard rodent pellets and water, the rats immediately gorged on the human food, losing interest in their boring pellets. On this diet, they gained weight at an unprecedented rate. In a few short weeks, the rats were obese, and neither exercise nor an enriched environment were able to prevent it (although exercise did attenuate it). Sclafani named this the "supermarket diet," although most researchers now call it the "cafeteria diet."

Sclafani's study was published in 1976, and to this day, the cafeteria diet remains the most effective way to get a normal rat or mouse to overeat—far more effective than diets that are simply high in fat and/or sugar.

This leads us to a disturbing conclusion: Palatable human food is the most effective way to cause a normal rat to spontaneously overeat and become obese, and its fattening effect cannot be attributed solely to its fat or sugar content. If that's true, then what does this food do to humans?

ONE WEIRD TRICK TO MAKE A HUMAN OVEREAT

In the early 1990s, Eric Ravussin, currently director of the Nutrition Obesity Research Center at the Pennington Biomedical Research Center in Baton Rouge, and his team were looking for a better way to measure calorie and nutrient intake in humans. This turns out to be a surprisingly difficult task. At the time, a number of studies had reported that people with obesity eat about the same number of calories as lean people, leading some researchers to question the role of calorie intake in obesity. The catch was that the data were all self-reported. In other words, researchers asked people to describe what they had eaten and then tallied up the resulting calories. This method does have its advantages; in particular, it provides a snapshot of what people eat in their everyday lives.

It also has disadvantages, which became apparent when researchers began to use more accurate methods of measuring calorie intake. These experiments indicated that people with obesity consistently eat more calories than lean people—after factoring in height, gender, and physical activity (as explained previously, the fact that they have to eat more calories to maintain weight is due to their larger tissue mass). This suggested that the self-reported calorie intake data were wrong in a very misleading way. Today, we know from a variety of lines of evidence that this is indeed the case: People are notoriously bad at describing what, and particularly how much, they eat. Ravussin knew he couldn't use this method if he wanted accurate results.

A more rigorous approach is to lock people inside a research facility called a metabolic ward and feed them a carefully controlled diet in which every morsel of food is precisely quantified by researchers. This is an extremely accurate way to measure food intake, but it's also unnatural. People don't get to choose what they eat, so their food intake may not reflect their normal eating habits. The results of these studies are very reliable but may lack real-world validity.

Ravussin and his team wanted an intermediate method, one that had the accuracy of the metabolic ward but allowed people to select their own diets, replicating everyday life as much as possible. Their solution was to install enormous U-Select-It 3007 vending machines inside a metabolic ward, each containing a variety of entrées, snacks, and beverages. The foods in the vending machines were not chosen at random—"We were screening what people liked and disliked," explains Ravussin—and only included foods that were tempting. The menu included things like french toast and sausages with syrup, chicken pot pie, chocolate-vanilla swirl Jell-O pudding, cheesecake, nacho cheese Doritos, M&M's, Shasta cola, and a few apples for the health nuts (no Froot Loops, sadly)—in other words, mostly "palatable human foods" similar to those Sclafani used in his rat study. Ten male volunteers were locked up with the vending machines for seven days with the freedom to select when and what they ate. To monitor their food intake, each person had to enter an ID code when he withdrew food and return all uneaten food to the research staff for weighing.

The experiment was a success: Ravussin's team was able to accurately measure food intake in people who were selecting their own food, and at the same time take a number of other informative metabolic measurements. Yet during the course of the study, something remarkable dawned on Ravussin: The volunteers were overeating tremendously. "People were on average eating almost double the amount they needed," he recalls. To be precise, the volunteers averaged 173 percent of their normal calorie needs, overeating from day one until the end of the experiment. Over the course of the seven-day study, the men gained an average of five pounds each.

Over the next three years, Ravussin's team published two additional "human cafeteria diet" studies. These studies tested the vending machine setup with men, women, lean people, people with obesity, Caucasians, and Native Americans. In each case, volunteers locked in a metabolic ward with a variety of free, tasty foods overate substantially, without being asked to overeat. Ravussin named this phenomenon *opportunistic voracity*.

These findings are particularly remarkable because it's normally quite difficult to get people to overeat substantially for more than a few days (imagine yourself eating twice as much food at each meal!). In other settings, researchers coax their volunteers to overeat using enticements such as money, and even then, volunteers have to force down the extra food against a growing sense of queasiness and impending stomach rupture. Yet in Ravussin's study, his volunteers overate cheerfully without even being asked to, suggesting that the environment he created had the peculiar ability to sweep aside our natural limits on eating.

ENTER THE BRAIN

Just as Yutala gained weight when he left the traditional diet and lifestyle of his native island of Kitava, Americans gained weight as our own way of life evolved. Our current food environment shares many parallels with Sclafani's and Ravussin's cafeteria diet studies. To understand why we overeat when we're placed in such environments, and why we do so without having any conscious intention or desire to overeat, we must turn to the organ that controls all behavior, including eating: the brain.

2

THE SELECTION PROBLEM

I n the basement of Sten Grillner's laboratory at the Karolinska Institute in Stockholm, Sweden, dozens of foot-long, wormlike creatures adhere to the glass wall of a large fish tank using round, suction-cup-like mouths filled with needle-sharp teeth. These nightmarish creatures are lampreys, one of our truly ancient relatives (see figure 6). Lampreys and their cousins the hagfish are considered the most primitive living vertebrates, meaning animals that have evolved a backbone, spinal cord, and brain.[16] The ancestors of today's lampreys diverged from our own ancestors approximately 560 million years ago—before the evolution of mammals, dinosaurs, reptiles, amphibians, and even most types of fishes—and long before our ancestors laid a fin on land.

Because lampreys are our most distant vertebrate relatives, comparing their brains to the brains of mammals reveals the common elements of all vertebrate brains: the core processing circuits that form the foundation of the human mind. Grillner's research has shown that within the pea-sized brain of these primitive creatures lie the seeds of the human

---

16 Technically, lampreys don't have a backbone (vertebral column), but they do have a spinal cord and brain, and they are thought to have had a backbone that they subsequently lost during their evolution.

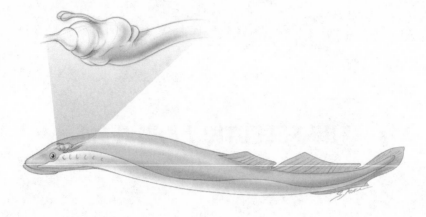

Figure 6. The river lamprey, *Lampetra fluviatilis,* and its brain.

decision-making apparatus.[17] If we are to fathom our own eating behavior, we have to understand the fundamentals of how the brain makes decisions, and the lamprey is an excellent place to start.

THE SELECTION PROBLEM: HOW DECISIONS ARE MADE IN A COMPLEX WORLD

Imagine two robots on a car assembly line. As each car door rolls by, Robot 1 paints it green. Door after door, Robot 1 performs the exact same action, and that single action is all it can do. This kind of robot doesn't require much processing power because it only has one job, one capability, and therefore no decisions to make. Now, imagine that Robot 2 can do two different things: It can paint a door green, or it can paint it red. Robot 2 only has one paint nozzle, so it can't paint a door both colors at once. Rather, it has to choose which color to use. So how does Robot 2 decide? This fundamental challenge is called the *selection problem,* and it occurs any time there are multiple options (green versus red paint) competing for the same shared resource (one paint nozzle). To solve the selection problem,

---

17 Incredibly, the very earliest roots of the basal ganglia probably predate vertebrates, because similar structures have been identified in the brains of flies. Our ancestors have been using their brains to make decisions for a *very* long time.

Robot 2 needs a *selector*—some sort of function that decides which color is most appropriate to apply to each door.

Our very earliest ancestors were probably like Robot 1—simple creatures that didn't have to make any decisions about what to do. But this didn't last long. As soon as they evolved the ability to do more than one thing with the same set of resources, they had to begin making decisions about what to do—and those that made the best decisions passed on their genes to the next generation.[18] Lampreys, for example, can do a number of different things: They can suction onto a rock, track prey, flee predators, mate, build a nest, feed, and swim in a nearly infinite number of directions. Many of these options are mutually exclusive because they require the same muscles inside the same body. So like Robot 2, the lamprey has a selection problem, and it needs a selector to resolve it.

According to researchers in computational neuroscience and artificial intelligence, an effective selector must have certain key properties, whether it exists in a computer or in a brain:

1. The selector must be able to choose *one* option. If there are incompatible options, such as fleeing a predator or mating, the selector must be able to pick only one and allow it access to the resources necessary to execute its program.
2. The selector must be able to choose the *best* option in any given situation. For example, if a lamprey sees a dangerous predator, it should flee.[19] A lamprey that tries to mate when it sees a dangerous predator won't survive and won't pass on its genes to the next generation of lampreys.
3. The selector must be able to select *decisively* between options. If one option is only slightly better than the others, it still must

---

18 The earliest "decisions" didn't require neurons or a brain, just as today's bacteria can make simple decisions. For example, many bacteria can move toward food sources and away from noxious chemicals, a behavior called *chemotaxis*. They're able to "decide" which direction is most appropriate to swim based on information about their surroundings.

19 To an adult lamprey, food means a fish it can latch on to and parasitize. The lamprey uses its round mouth full of sharp teeth to rasp away the flesh of its host, which often results in the untimely death of the fish. I told you they're nightmarish!

win definitively, shutting off all incompatible options completely. A lamprey that tries to mate and flee at the same time is probably not going to leave many offspring.

In a seminal research paper published in 1999, University of Sheffield researchers brought together evidence from neuroscience and computer modeling to argue that selection is precisely the function of an ancient group of structures deep within the human brain called the *basal ganglia*. Today, this idea is accepted by most neuroscientists. To understand how the human selector works, let's start with a simpler version of it: the lamprey selector.

THE LAMPREY SOLUTION TO THE SELECTION PROBLEM

How does the lamprey decide what to do? Within the lamprey basal ganglia lies a key structure called the *striatum,* which is the portion of the basal ganglia that receives most of the incoming signals from other parts of the brain.[20] The striatum receives "bids" from other brain regions, each of which represents a specific action. A little piece of the lamprey's brain is whispering, "Mate," to the striatum, while another piece is shouting, "Flee the predator!" and so on. It would be a very bad idea for these movements to occur simultaneously—because a lamprey can't do them all at the same time—so to prevent simultaneous activation of many different movements, all these regions are held in check by powerful inhibitory connections from the basal ganglia.[21] This means that the basal ganglia keep all behaviors in "off" mode by default. Only once a specific action's bid has been selected do the basal ganglia turn off this inhibitory control, allowing the behavior to occur (see figure 7). You can think of the basal ganglia as a bouncer that chooses which behavior gets access to the muscles and turns away the rest.

---

20 The striatum is often divided into two sections: the dorsal (upper) striatum, and the ventral (lower) striatum, which is also commonly called the nucleus accumbens. These play different roles in selection, a topic to which we'll return.

21 Parts of the globus pallidus and substantia nigra.

This fulfills the first key property of a selector: It must be able to pick *one* option and allow it access to the muscles.

Many of these action bids originate from a region of the lamprey brain called the *pallium,* which is thought to be involved in planning behavior. Each little region of the pallium is responsible for a particular behavior, such as tracking prey, suctioning onto a rock, or fleeing predators.

These regions are thought to have two basic functions. The first is to execute the behavior in which it specializes, once it has received permission from the basal ganglia. For example, the "track prey" region activates downstream pathways that contract the lamprey's muscles in a pattern that causes the animal to track its prey.

The second basic function of these regions is to collect relevant information about the lamprey's surroundings and internal state, which determines how strong of a bid it will put in to the striatum[22] (figure 7). For example, if there's a predator nearby, the "flee predator" region will put in a very strong bid to the striatum, while the "build a nest" bid will be weak. If the lamprey is hungry and it sees prey, the "track prey" bid will be strong, but the "suction onto a rock" bid will be weak.

Each little region of the pallium is attempting to execute its specific behavior and competing against all other regions that are incompatible with it. The strength of each bid represents how valuable that specific behavior appears to the organism at that particular moment, and the striatum's job is simple: Select the strongest bid. This fulfills the second key property of a selector—that it must be able to choose the *best* option for a given situation.

At the same time that the striatum selects the strongest bid, it shuts down competing bids. So once the "flee predator" bid has won the competition, bids like "suction onto a rock" and "track prey" get bounced. This fulfills the third key property of a selector—that it must be able to select *decisively,* picking one option and shutting down all competitors.

Each region of the pallium sends a connection to a particular region of the striatum, which (via other parts of the basal ganglia) returns a connection back to the same starting location in the pallium. This means that

---

22 The strength of a bid is represented by the strength of neuron firing entering the striatum. "In the brain," says Grillner's former graduate student Marcus Stephenson-Jones, "valuation is neural firing."

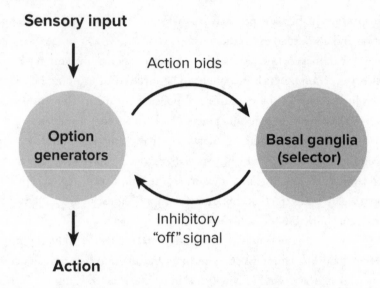

Figure 7. A general model of action selection by the basal ganglia. Adapted from McHaffie et al., *Trends in Neurosciences* 28 (2005): 401.

each region of the pallium is reciprocally connected with the striatum via a specific loop that regulates a particular action (see figure 7). For example, there's a loop for tracking prey, a loop for fleeing predators, a loop for anchoring to a rock, and so on. Each region of the pallium is constantly whispering to the striatum to let it trigger its behavior, and the striatum always says "no!" by default. In the appropriate situation, the region's whisper becomes a shout, and the striatum allows it to use the muscles to execute its action. This is how the lamprey is able to react appropriately to its surroundings and internal state.[23]

With all this in mind, it's helpful to think of each individual region of the lamprey pallium as an *option generator* that's responsible for a specific behavior. Each option generator is constantly competing with all other incompatible option generators for access to the muscles, and the option

---

23 The basal ganglia operate in a remarkably similar manner to decision-making systems independently designed by engineers, which pit competing options against one another to make optimal decisions under complex conditions. This suggests that this competitive decision-making strategy may be a universally optimal way of making decisions.

generator with the strongest bid at any particular moment wins the competition. The basal ganglia evaluate the bids, decide which one is the strongest, give the winning option generator access to the muscles, and shut down all competing option generators (see figure 7). The lamprey flees, avoids the predator, and lives to pass on its genes to the next generation of lampreys.

THE MAMMALIAN SOLUTION TO THE SELECTION PROBLEM

Most people would agree that the human brain is a bit more sophisticated than the brain of a lamprey. All right, *a lot* more sophisticated. One of the things that sets mammals apart from most other creatures on earth is the tremendous complexity of our nervous systems, which allows us to make remarkably smart decisions. To understand just how useful our high-powered model is, consider how much energy it guzzles. In humans, the brain eats up *one-fifth* of our total energy usage, even though it accounts for only *2 percent* of our body weight. The fact that evolution allowed us to bear this energy-hogging ball and chain is a testament to its importance. Making smart decisions is a good evolutionary strategy, and no animal does it better than humans.

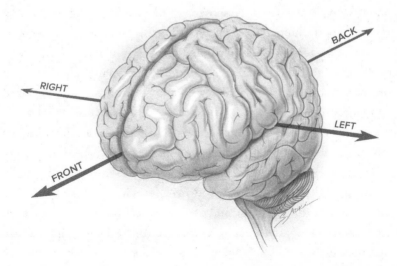

Figure 8. The human brain.

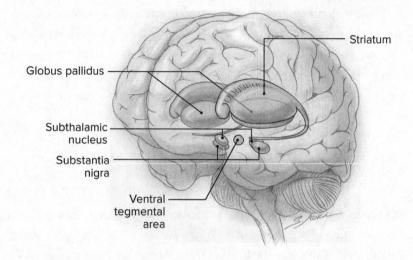

Figure 9. The human basal ganglia, with nuclei labeled. The striatum is composed of two nuclei called the caudate nucleus and the putamen.

So what does the lamprey brain have to do with the human brain? This is what Karolinska Institute researcher Sten Grillner and his former graduate student Marcus Stephenson-Jones set out to answer. Building on the work of previous researchers, they compared the anatomy and function of the basal ganglia in lampreys and mammals (see figure 9 for an illustration of the human basal ganglia). Their findings are nothing short of remarkable: Despite being separated by a 560-million-year evolutionary chasm, the basal ganglia of lampreys and mammals (including humans) are strikingly similar. They contain the same regions, organized and connected in the same way. Within these regions lie neurons with the same electrical properties, communicating with one another using the same chemical messengers. These findings led Grillner and Stephenson-Jones to the stunning conclusion that "practically all details of the basal ganglia circuitry had developed some 560 million years ago." Stephenson-Jones adds: "This is really a fundamental part of the vertebrate brain that's been used across evolution as a common mechanism for how lampreys, fish, birds, mammals, and even humans make decisions." Our ancestors hit an evolutionary home run 560 million years ago, and we still carry the hardware they developed in the ancient seas.

A lamprey is capable of making many different types of decisions, but certainly far fewer than humans are. We have to make up our minds about things a lamprey cannot fathom, like what to cook for dinner, how to pay off the mortgage, and whether or not to believe in God. Clearly, there are important differences in the brain hardware that allows us to understand the world and make choices. But if the decision-making capacity of a human and a lamprey are so different, why are the basal ganglia of lampreys and humans so strikingly similar? Grillner and Stephenson-Jones propose an explanation: an evolutionary process called *exaptation*. As opposed to *adaptation,* which is the process of developing new traits—such as air-breathing lungs or a four-chambered heart—exaptation takes something that already exists and finds a new function for it; for example, expanding the basal ganglia's decision-making jurisdiction to govern other, more advanced types of decisions. Grillner and Stephenson-Jones propose that the early vertebrate basal ganglia were already very good at making decisions, and there was no need for evolution to fix what wasn't broken. We only needed to build on it.

In humans, the most numerous inputs to the striatum come from the *cerebral cortex,* which evolved from the pallium (similar to the one found in today's lampreys). The cortex is critical for advanced decision-making. You can still do a lot of basic things without it, things that are regulated by deeper, older parts of the brain,[24] but you can't decide about the mortgage or about God. The cortex is comically enlarged in humans relative to other animals, and it plays a key role in our exceptional intelligence. The lamprey pallium is rudimentary by comparison (see figure 10).[25] This is part of the reason why lampreys don't have mortgages.

These major inputs from the cortex to the striatum suggest that the role of the basal ganglia has expanded considerably since our divergence from the ancestors of lampreys. As it turns out, the cortex doesn't just send inputs to the basal ganglia—it also receives input back, just like the lamprey pallium.[26] These reciprocal connections also form loops that

---

24 Mammals such as rats retain most basic aspects of normal behavior even when the cortex has been completely removed! For example, they can still eat, walk, mate, and learn simple tasks. These animals run into trouble when they have to act in more sophisticated, flexible ways.

25 The lamprey pallium is tiny compared to the mammalian cortex, and it doesn't have the same cellular organization.

26 These connections are relayed to the cortex using the thalamus as an intermediary.

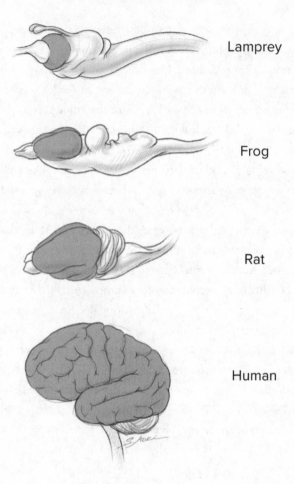

Lamprey

Frog

Rat

Human

Figure 10. Lamprey, frog, rat, and human brain, with cortex (or pallium) highlighted.

travel to and from specific regions of the cortex, each of which is an option generator. In fact, there are similar loops connecting the basal ganglia to many parts of the mammalian brain—parts that regulate not only movements but motivations and emotions, thoughts and associations, and numerous other processes.

Over the course of evolutionary history, the process of exaptation multiplied the basic decision-making units of the basal ganglia and connected them to fancy new option generators capable of proposing much more sophisticated options and computing value in more advanced ways. In addition to deciding how to move, the human basal ganglia can decide how to feel, what to think, what to say, and—to return to the matter at hand—what to eat.

THE BASAL GANGLIA GO TO A RESTAURANT

Broken down to its fundamental elements, behavior begins to appear quite complex because it involves the coordination of many interacting parts. To achieve a seemingly simple goal such as eating at a restaurant, you must first become motivated to eat, then you must figure out where you want to eat and how to get there, and then you must control your musculature in just the right way to get there and place the food into your mouth. This is vastly more challenging than the task that Robot 2 faced, because each of these processes involves a decision. These motivational, cognitive, and motor tasks are all processed in different parts of the brain, yet they work together so seamlessly that we're scarcely aware they're distinct from one another. How does the brain make all these decisions in such a coordinated manner?

It's impossible to be certain, because we can't do detailed invasive studies on the human brain as we do in other species, but researchers have a compelling hypothesis that draws from a variety of scientific clues. To understand this hypothesis, I spoke with University of Sheffield researchers Peter Redgrave and Kevin Gurney, who have played key roles in uncovering the function of the basal ganglia in decision-making. Here's what they explained to me:

Let's say you haven't eaten in a while. From a survival standpoint, your body wants energy, so eating would be a valuable action. How does it get you to do it? The first step is to set your motivation for food. The ventral (bottom) part of the striatum is responsible for selecting between competing motivations and emotions.[27] "Those are the motivational channels that select high-level goals," explains Redgrave. "Are you hungry, thirsty, frightened, sexy, cold, or hot?" The hungry, thirsty, frightened, sexy, cold, and hot option generators compete for expression by sending bids to the ventral striatum. At the moment, because you're low on energy, the hunger option generator is putting in a very strong bid (a potential pitfall we'll explore later on). It wins the competition and is allowed to express itself. You begin to feel hungry.

Once the hunger option generator has won and you feel motivated to

---

27 Also called the nucleus accumbens, an infamous brain region we'll come back to in the next chapter. It's worth noting that the functions of the ventral and dorsal striatum overlap somewhat, although for simplicity's sake I'll be treating them as distinct in this book.

eat, it begins to activate other option generators in the cortex responsible for figuring out how to get food, or generally, making a plan. Option generators representing the refrigerator, pizza delivery, the restaurant down the street, and the really good restaurant across town enter into competition in the dorsal (upper) striatum. Just as the strength of the hunger bid was determined by information about your body's energy status, the strength of each eating plan bid is determined by relevant information about that option: how good the food was last time you had it, what other people have said about it, how much effort it requires, and how expensive it is. As it turns out, the restaurant across town is really good, but you don't feel like driving. The food in your fridge is cheapest, but it requires cooking. The restaurant up the street is close and cheap, and so it puts in the strongest bid and wins the competition.

Now you have a plan, but how do you execute it? Do you walk, ride your bike, drive, or take the bus? The "restaurant up the street" option generator initiates another competition between walk, bike, and bus option generators in the cortex, which again provide competing inputs to the dorsal striatum. You feel like getting some fresh air, but you also want to get there quickly, so the bike option wins. Once you get onto your bike, how do you make it move forward? Do you wave your hands, wiggle your toes, shake your head, or pedal? The answer is obvious, but it still requires a decision between competing options in motor regions of the brain. Pedaling puts in the strongest bid, and you jump on your bike and head for the restaurant. This process is illustrated schematically in figure 11, and the underlying brain circuits are illustrated in figure 12.

I'll spare you the description of all the decisions you have to make while riding to the restaurant, choosing from the menu, and eating your food. The point is that many of our behaviors are thought to result from a cascading series of competitions within motivational, cognitive, and motor brain regions. The winning motivation initiates subsequent competitions in cognitive areas that are relevant to fulfilling that motivation, and then the cognitive areas initiate competitions in motor areas that are relevant to physically executing the plan of action. The strength of each bid is determined by experience, internal cues, and external cues, and the basal ganglia only allow the strongest bids to express themselves. This process occurs beyond our conscious awareness—we only become aware of bids after

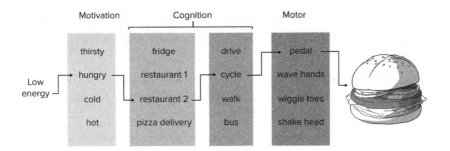

Figure 11. A cascading decision-making process leading to food consumption. First, the brain senses low energy stores, causing the "hunger" option generator to outbid other possible motivations. Then the "hunger" option generator activates cognitive regions that compete to decide how to get food, and then the cognitive option generators activate relevant motor regions that compete to execute the appropriate movements.

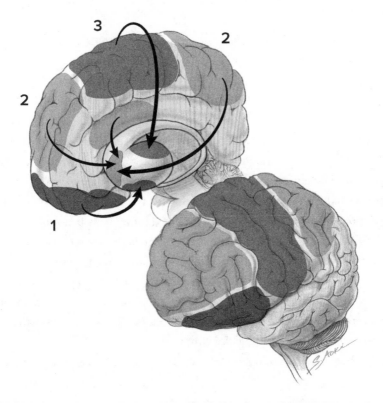

Figure 12. How a decision-making process may happen in the brain. First, regions of the prefrontal cortex send bids to the ventral striatum to select a goal. Second, cognitive areas of the cortex send bids to the medial striatum to select a plan. And third, motor areas of the cortex send bids to the dorsal striatum to select movements.

they're selected.[28] This is consistent with Daniel Kahneman's idea (discussed in the introduction) that most of what happens in the brain, including many decision-making processes, is nonconscious.

Many behaviors we view as trivial, such as pumping gasoline or washing the dishes, are actually tremendously complex. Artificial intelligence researchers are keenly aware of the difficulties of reproducing even elementary goal-directed behaviors, which explains why today's computers are good at computing but not so good at making complex decisions without human guidance. This is a testament to how much of the functioning of our own brains we take for granted.

THE MAN WHO HAD NO THOUGHTS

To illustrate the crucial importance of the basal ganglia in decision-making processes, let's consider what happens when they *don't* work.

As it turns out, several disorders affect the basal ganglia. The most common is Parkinson's disease, which results from the progressive loss of cells in a part of the basal ganglia called the *substantia nigra*. These cells send connections to the dorsal striatum, where they produce *dopamine,* a chemical messenger that plays a very important role in the function of the striatum. Dopamine is a fascinating and widely misunderstood molecule that we'll discuss further in the next chapter, but for now, its most relevant function is to increase the likelihood of engaging in any behavior.

When dopamine levels in the striatum are increased—for example, by cocaine or amphetamine—mice (and humans) tend to move around a lot. High levels of dopamine essentially make the basal ganglia more sensitive to incoming bids, lowering the threshold for activating movements. Figure 13 illustrates the striking effect of dopamine-boosting cocaine on mouse locomotion (walking and running).

Conversely, when dopamine levels are low, the basal ganglia become

---

28 Many neuroscientists believe we become aware of decisions after they've already been made by unconscious parts of the brain. Redgrave: "You do realize with these ideas there is a really, really *horrendous* implication. We become aware of what is selected *after* it's been selected, not before. You have all these competing things, one wins, and you become aware of that winner." The "horrendous implication" is that we can't have free will if our decisions are made unconsciously.

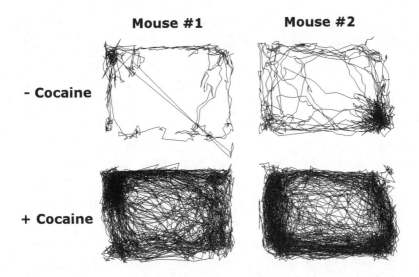

Figure 13. The effect of cocaine on locomotion in mice. Each line represents the movement path of one mouse inside its cage over a twenty-minute period. Two mice were injected with a saline solution (–cocaine; upper traces) and then, two days later, with saline plus cocaine (+cocaine; lower traces). Note the marked increase in locomotion in the cocaine-injected mice. Images courtesy of Ross McDevitt, US National Institute on Drug Abuse.

less sensitive to incoming bids and the threshold for activating movements is high. In this scenario, animals tend to stay put. The most extreme example of this is the dopamine-deficient mice created by Richard Palmiter, a neuroscience researcher at the University of Washington. These animals sit in their cages nearly motionless all day due to a complete absence of dopamine. "If you set a dopamine-deficient mouse on a table," explains Palmiter, "it will just sit there and look at you. It's totally apathetic." When Palmiter's team chemically replaces the mice's dopamine, they eat, drink, and run around like mad until the dopamine is gone.

In Parkinson's disease, the gradual loss of substantia nigra neurons causes dopamine levels to decline in the areas of the dorsal striatum that select movements, particularly well-worn movement patterns. This makes the dorsal striatum progressively less sensitive to incoming bids from motor regions, rendering it increasingly difficult for any motor option generator to gain access to the body's muscles. People with Parkinson's disease develop difficulty initiating and executing movements and have a particularly hard time performing sequences of movements. In severe cases, Parkinson's disease patients are scarcely able to initiate movements at all, a phenomenon called akinesia, Greek for "no movement."

Fortunately, modern medicine has developed drugs that help relieve the debilitating motor impairments of Parkinson's disease. Most of these are designed to increase dopamine signaling in the brain. The most effective and commonly used drug is the dopamine precursor L-dopa. When taken orally, L-dopa enters the circulation, and some of it crosses into the brain. Once inside the brain, it's taken up by dopamine-producing neurons and converted into dopamine. Although there's currently no way to regenerate lost cells in the substantia nigra, L-dopa makes the remaining cells, and perhaps even other cell types that don't normally contain dopamine, produce more dopamine to compensate for the deficit.[29] Higher levels of dopamine make the dorsal striatum more sensitive to incoming bids from motor regions and enables patients to move more normally once again.

As with many drugs, L-dopa is a blunt tool. In Parkinson's disease, parts of the dorsal striatum need more dopamine—but the rest of the brain doesn't. When a person takes L-dopa, dopamine-producing neurons throughout the brain—including those located in the area that provides dopamine to the ventral striatum, the ventral tegmental area (VTA)—sponge it up and convert it to dopamine. This can lead to abnormally elevated levels of dopamine in the ventral striatum.

As I mentioned earlier, the ventral striatum primarily regulates motivations and emotional states. Analogous to what happens in the dorsal striatum, elevating dopamine in the ventral striatum makes it more sensitive to incoming bids, increasing the likelihood that it will activate motivational and emotional states. In fact, common side effects of L-dopa treatment include heightened emotional states, hypersexuality, and compulsive and addictive behaviors, such as gambling, shopping, drug abuse, and binge eating. These are called *impulse control disorders* because people lose the ability to keep their basic impulses in check. The ventral striatum is so sensitive to incoming bids that inappropriate option generators are able to grab the reins. In addition, higher levels of dopamine in the striatum may cause the

---

29 Redgrave notes that other neurons (e.g., those that normally produce serotonin) may also participate in taking up L-dopa, converting it into dopamine, releasing it into the striatum, and alleviating Parkinson's disease.

activity of certain loops to become abnormally strong over time, resulting in addictive and compulsive behaviors, a topic we'll discuss in the next chapter.

Other basal ganglia disorders are even more interesting. Consider Jim, a former miner who was admitted to a psychiatric hospital at the age of fifty-seven with a cluster of unusual symptoms.[30] As recorded in his case report:

> During the preceding three years he had become increasingly withdrawn and unspontaneous. In the month before admission he had deteriorated to the point where he was doubly incontinent, answered only yes or no to questions, and would sit or stand unmoving if not prompted. He only ate with prompting, and would sometimes continue putting spoon to mouth, sometimes for as long as two minutes after his plate was empty. Similarly he would flush the toilet repeatedly until asked to stop.

Jim was suffering from a rare disorder called abulia, which is Greek for "an absence of will."[31] Patients who suffer from abulia can respond to questions and perform specific tasks if prompted, but they have difficulty spontaneously initiating motivations, emotions, and thoughts. A severely abulic patient seated in a bare room by himself will remain immobile until someone enters the room. If asked what he was thinking or feeling, he'll reply, "Nothing." Needless to say, abulic patients have little motivation to eat.

Abulia is typically associated with damage to the basal ganglia and related circuits,[32] and it often responds well to drugs that increase dopamine signaling. One of these is bromocriptine, the drug used to treat Jim:

> He was started on bromocriptine 5 mg per day, and increased in 5 mg stages to a maximum of 55 mg in divided doses. His first spontaneous activity, dressing without prompting, occurred at 20 mg. At 30 mg he started to initiate conversations with other residents, but there was considerable

---

30 Jim's real name is not provided in the case report.

31 Also called "psychic akinesia."

32 Carbon monoxide poisoning is often the cause of this damage; the basal ganglia appear to be particularly sensitive to it.

day to day variation. As the dose increased, he washed, dressed and ate his meals without prompting and was not perseverative. On some days though, he reverted to his pre-treatment state. On the maximum dosage such days were rare and he was generally independent in his activities of daily living . . .

Researchers believe that the brain damage associated with abulia causes the basal ganglia to become insensitive to incoming bids, such that even the most appropriate feelings, thoughts, and motivations aren't able to be expressed (or even to enter consciousness). Drugs that increase dopamine signaling make the striatum more sensitive to bids, allowing some abulic patients to recover the ability to feel, think, and move spontaneously.

WHAT DOES THIS HAVE TO DO WITH OVEREATING?

Now that we know something about how the brain makes decisions, we can get more specific about how it decides what, and how much, to eat. Eating is a complex behavior that requires coordinated decision-making on motivational, cognitive, and motor levels. The fundamental spark that sets the whole behavioral cascade in motion, however, is *motivation*. This motivation to eat can come from several different brain regions, in response to different cues. For example, the option generator that causes hunger is presumably different from the one that causes you to eat dessert after a big meal, and is also presumably different from the one that caused Joey Chestnut to eat sixty-nine hot dogs in ten minutes to win Nathan's Hot Dog Eating Contest. Yet without motivation, there is no eating.

In the next chapter, we'll delve into the circuits that set our motivation for eating food—particularly those that drive us to overeat. Which brain circuits motivate us to overeat, what cues drive that motivation, and what can we do about it? We'll start off by sticking with the basal ganglia, exploring how these structures cause us to learn about food, crave it, and perhaps even become addicted to it.

3

THE CHEMISTRY OF SEDUCTION

You have just emerged from your mother's womb into a hospital room full of strangers, bright lights, and machines. Bewildered by the multitude of new sights and sensations, you begin to cry. At this point, crying is one of the few things you know how to do, part of a small repertoire of instinctive behaviors like suckling. Over the course of your life, however, you'll develop the desire and ability to play with blocks, read written words, hit a baseball, kiss another person, hold a job, and acquire and eat everyday foods. This striking behavioral transformation is due to a phenomenon, often taken for granted, called *learning*. Learning is the process of acquiring new knowledge, skills, movement patterns, motivations, and preferences, or reinforcing those that already exist. As it turns out, learning—particularly the effects of learning on our motivation to seek certain foods—is one of the key reasons we overeat, despite our better judgment.

In order to learn, you must start with a goal. If you don't have a goal, you can't determine which behaviors are more valuable than others and, therefore, which should be cultivated. In the evolutionary sense, the ultimate goal of any organism is to maximize its reproductive success: have as many high-quality offspring as possible, which in turn will produce as many high-quality offspring as possible.[33] But that's not the goal we're thinking about when we dig into a bowl of cereal; in fact, we're rarely, if ever, aware of it. What we *are* aware of is a variety of proximate goals that natural selection has hardwired into our brains as a shorthand for the ultimate goal of reproductive success. For most animals, these goals include obtaining food and water, mating, seeking safety from danger, and seeking physical comfort. Humans, being more complex and social than most other animals, also seek social status and material wealth (although we can't lay exclusive claim to this; many social animals, including chimpanzees, use favors, sex, and violence to climb the social ladder). These goals—to eat, to drink, to have sex, to be safe and comfortable, to be liked—are the fundamental drivers of motivation and learning. Because food is so important for survival and reproduction, it tends to be a very powerful teacher.

When we hear the word *learning,* we tend to imagine ourselves poring over a textbook absorbing facts, but nearly everything we do—and think and feel—was learned at some point, whether intentionally or not. Roy Wise, a motivation and addiction researcher at the National Institute on Drug Abuse in Baltimore, Maryland, made this point in a 2004 review paper:

> Most goal-directed motivation—even the seeking of food or water when hungry or thirsty—is learned. It is largely through selective reinforcement of initially random movements that the behavior of the [newborn] comes to be both directed and motivated by appropriate stimuli in the environment.

To give you an example, imagine an infant trying to grab the tail of a cat sitting in front of him. He's not coordinated enough, and he mostly flails around, sometimes touching the tail but having a hard time grasping it. Suddenly, by chance, his arm and hand move in just the right way, and he grabs the cat's tail for a brief moment. Realizing that something good just happened, his brain increases the likelihood that the same movement

---

33 Richard Dawkins and others have argued, convincingly, that the primary unit of natural selection is the gene rather than the whole organism. For our purposes, referring to the organism as the unit of selection is close enough and much more intuitive. Readers who would like to delve deeper into this topic are referred to Dawkins's book *The Selfish Gene.*

pattern will recur the next time he wants to grab the cat's tail. With practice, his brain refines the movement, until he can terrorize the cat at will. More generally, when something good happens, the brain increases the likelihood that the pattern of brain activity that immediately preceded the good event will recur in the future. To put this in terms of what we covered in the last chapter, the successful option generators get a stronger bid.

From the outside, we observe that when a behavior meets a goal, it is more likely to recur in the future—it's *reinforced*. The noted American psychologist Edward Thorndike described the phenomenon of reinforcement as early as 1905, stating that "any act which in a given situation produces satisfaction becomes associated with that situation so that when the situation recurs the act is more likely than before to recur also." Over the course of our lives, with experience, we refine our ability to achieve our goals, and reinforcement is one of the simplest and most powerful means of doing so.

To further illustrate this, let's return to the example of getting yourself something to eat. To satisfy your hunger, you activated option generators representing the restaurant up the street, getting on your bike, and pedaling. This is the pattern of motivations, thoughts, and movements that got you to the restaurant. Now, let's say you ate at the restaurant and the food was *really* good—unexpectedly good. You achieved your food goal very effectively. The option generators that got you to the restaurant will be strengthened, and the next time you're hungry, you'll be more likely to crave the restaurant up the street, and perhaps even hop on your bike to get there. You'll come to enjoy the thought, appearance, and smell of the restaurant. Your behavior of going to the restaurant up the street will be reinforced.

Learning shapes all three levels of our decision-making process: motivational, cognitive, and motor. Reinforcement strengthens them all, because they're all required for effective goal-directed behaviors. The process of reinforcement operates completely outside our conscious awareness, and it has existed since before the time of our common ancestor with lampreys.

Learning also operates in the opposite direction. When something bad happens as the result of a behavior, the likelihood of that behavior recurring decreases. For example, if you develop food poisoning after eating at the restaurant up the street, you'll be less likely to eat there again. You won't crave the restaurant when you're hungry, and the thought, appearance, and smell of the restaurant may even provoke a feeling of disgust. This is called *negative reinforcement.*

For reinforcement to occur, there has to be a teaching signal that changes the activity of basal ganglia loops based on experience, such that good responses are reinforced and bad responses are discarded. Most researchers believe the brain's teaching signal is the fascinating molecule dopamine.[34]

THE LEARNING CHEMICAL

Ross McDevitt, a postdoctoral researcher at the National Institutes of Health in Baltimore, Maryland, gently places a mouse into a clear plastic cage and presses a thin fiber-optic cable into a tiny connector on the animal's head. McDevitt is using a cutting-edge technique called optogenetics to specifically stimulate cells in the mouse's ventral tegmental area (VTA). As we discussed in the last chapter, the VTA is a brain region that sends dopamine-laden fibers to the primary motivation center of the brain: the ventral striatum (see figure 14). When these fibers release dopamine, they change the activity of cells in the ventral striatum and related brain regions, with profound consequences for behavior. Previously, we saw that high overall levels of dopamine can increase the likelihood that any option generator will grab the reins of behavior—yet it has other effects that are far more elegant. Dopamine, as a matter of fact, is the essence of reinforcement.

The end result of McDevitt's experimental setup is that he can send bursts of dopamine into the ventral striatum with the flick of a switch, which will illustrate the remarkable power of this pathway in learning and motivation.

The mouse's cage has a little box built into it, and each time the mouse

---

34 A small number of researchers don't think dopamine is a teaching signal. Kent Berridge is a notable example. The dopamine reinforcement debate is beyond the scope of this book, but scientifically proficient readers who would like an alternative perspective are encouraged to read Berridge's 2007 *Psychopharmacology* paper "The Debate over Dopamine's Role in Reward."

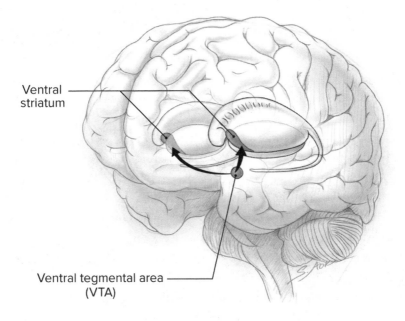

Ventral striatum

Ventral tegmental area (VTA)

Figure 14. The dopamine-releasing connection between the VTA and the ventral striatum. This pathway is central to reinforcement learning and motivation.

pokes its nose into the box, light travels down the fiber-optic cable on its head, shines onto its VTA neurons, and causes them to release bursts of dopamine into the ventral striatum and related brain regions. Yet the mouse doesn't know any of this at the beginning of the experiment. When the mouse is first placed into its cage, the little box is utterly meaningless. The mouse has no particular affinity for the box, and it pokes its nose into it only occasionally, out of curiosity. Each time the mouse nose-pokes, however, it experiences the mouse equivalent of eating chocolate, having sex, and winning the lottery all at the same time.

It doesn't take long for the nose-poking behavior to become more frequent. And more frequent. "What we find," explains McDevitt, "is that the mice go crazy for it. They love it." Whereas initially the mouse nose-pokes only out of curiosity, eventually it comes to understand the remarkable significance of the little box. McDevitt's mice end up nose-poking eight hundred times per hour—ignoring practically everything else in their cages. Other labs have shown that rats will nose-poke up to five thousand times per hour for VTA stimulation, which is more than once per

second! In other words, dopamine in the ventral striatum is highly reinforcing.

On a cellular level, this happens because dopamine acts on basal ganglia loops that were recently active, increasing the likelihood that they will
be activated again in the future. So whatever you're doing when the dopamine hits, you're more likely to repeat it when the same situation arises
again. The VTA basically says, "I like what just happened; I'm going to
sprinkle some dopamine into the ventral striatum to make sure it happens
again next time."

Although McDevitt evoked an exaggerated form of reinforcement by
directly stimulating the VTA, this process happens naturally in our brains
every day. When you accomplish a goal like eating a triple bacon cheese-
burger, dopamine is released in short bursts that reinforce your "successful" behavior. This is how dopamine teaches us how to feel, think, and
behave in ways that help us achieve our hardwired goals—whether or not
our conscious, rational brains support them. Dopamine in the ventral striatum is particularly important for learning motivations: for example,
learning which foods to crave and which to avoid.

Although dopamine wasn't discovered until a half century after his
initial experiments, the Russian physiologist Ivan Pavlov was one of the
first researchers to describe how animals learn to associate neutral cues
with food. Pavlov's team studied digestion in dogs, and he quickly learned
that his experimental animals drooled when they saw food—an observation that many dog owners can corroborate. Pavlov also noticed, with
some irritation, that the dogs would salivate even if he didn't have any
food. They had learned to associate Pavlov's presence with food.

Later, Pavlov's team found that when they consistently rang a bell prior
to feeding the dogs, the dogs eventually salivated in response to the bell
alone. The dogs had learned to associate the sound of the bell with the delivery of food, and because of this, the previously neutral sound cue had acquired importance. This is the same process that caused McDevitt's mice to
learn the importance of the initially boring nose-poke box as it was repeatedly associated with a powerful reward (dopamine in the ventral striatum).

Today, we call this process *Pavlovian conditioning,* and it's the reason
why the sight of cola on television, a little taste of ice cream, or the smell of
french fries can cause us to crave—and also to salivate. After you've learned

DOPAMINE: THE PLEASURE CHEMICAL?

You may have heard that dopamine is the "pleasure chemical," responsible for causing the neurochemical rush that makes us feel really good as we win a race, have sex, eat chocolate, or smoke crack cocaine. Although that idea is common in popular science writing, it has long been out of date within the scientific community. As a matter of fact, dopamine release doesn't correspond very well with the experience of pleasure. Experiments have shown that animals appear to experience pleasure without dopamine, and studies in humans back this up. Pleasure is more related to a class of chemicals called endorphins that are often released in the striatum simultaneously with dopamine, although these are probably only one component of the experience of pleasure. Dopamine is the "learning chemical" rather than the "pleasure chemical."

that the sight and smell of french fries predicts a fatty, starchy reward in your belly (with the help of dopamine), these sensory cues acquire importance and stimulate your motivation to eat french fries when you encounter them. Yet not all foods have this seductive effect on us. Why not?

THE CALORIE-SEEKING BRAIN

At this point, we know food can be a powerful reinforcer, tugging on the reins of our behavior. But some foods do this more than others. Brussels sprouts, for example, are a lot less seductive than ice cream. To understand why we overeat, we need to first answer the fundamental question: What is it about food, exactly, that's reinforcing? Anthony Sclafani, the researcher who invented the "cafeteria diet" we encountered in chapter 1, has dedicated the majority of his career to this question, and he's made remarkable progress.

Your average lab rat likes cherry-flavored water about as much as grape-flavored water. So if you place a bottle of each into its cage, it will drink about the same amount from each. However, in a groundbreaking study published in 1988, Sclafani's team showed that when they infused partially digested starch directly into the rats' stomachs while they drank cherry-flavored water, the rats developed a preference for that flavor over the grape flavor.[35] And the opposite preference developed when they repeated it with the grape flavor. Even though the starch never entered their mouths, after four days, the rats displayed a near-total preference for the starch-paired flavor. Sclafani called this phenomenon *conditioned flavor preference*.[36]

Conditioned flavor preferences are a remarkable phenomenon. The rats were totally unaware that researchers were infusing starch directly into their stomachs, yet somehow this starch sent a signal to the brain that caused the rats to increase their preference for a flavor they detected concurrently. In essence, they *learned* to prefer that previously neutral flavor just like Pavlov's dogs learned to salivate at the sound of a bell. How did this happen?

Further experiments showed that the rats weren't detecting the starch itself but the sugar glucose that's released as starch is broken up in the digestive tract. And the critical location for detection was the upper small intestine. Somehow, the intestine was sensing the glucose and sending a signal to the brain that said, "Something good just happened. Do that again!"

Sclafani still isn't certain how the signal gets from the gut to the brain, but he and other researchers have already figured out how the signal influences flavor preferences once it gets there.[37] The obvious suspect is—you guessed it—dopamine in the ventral striatum. Ivan de Araujo, associate professor of psychiatry at the Yale School of Medicine, and colleagues demonstrated independently of Sclafani that calories infused into the small intestine increase dopamine levels in the ventral striatum, and the more calories they infuse, the more dopamine spikes. Consistent with de Araujo's findings, Sclafani's group found that blocking dopamine action in the

---

35 Polycose, a type of maltodextrin.

36 I'm using the phrase *conditioned flavor preference* because it's the established scientific term; however, typically, the conditioned "flavor" is an odor. In other words, it's detected by olfactory receptors in the nose rather than taste receptors on the tongue.

37 The obvious suspect is the nerve fibers that connect the gut to the brain. Yet Sclafani's experiments to date haven't supported the hypothesis that they carry the signal. Sclafani speculates that it may be carried by a circulating hormone.

ventral striatum prevents the development of conditioned flavor preferences. "This suggests," says Sclafani, "that dopamine may be a central component of the process."

This research paints a nearly complete picture of how carbohydrate conditions flavor preferences. As a rat eats food, its mouth and nose detect the flavors and aromas associated with the food. After the rat swallows the food, it enters the rat's stomach and then its small intestine. The small intestine detects glucose and sends an unknown signal to the brain that causes dopamine to spike in the ventral striatum. If the food is rich in starch or sugar, a large spike in dopamine causes the rat to increase its preference for the flavors and aromas of the food it just ate—and become more motivated to seek foods with those flavors and aromas in the future. In this way, the rat becomes better at identifying and seeking foods that contain carbohydrate.

Sclafani's team has also been able to condition flavor preferences using fat and protein, demonstrating that rats respond to all three major classes of calorie-containing nutrients: carbohydrate (starches and sugars), fat, and protein.[38] Sclafani's work also revealed that the more concentrated the caloric load of a food—or the higher its *calorie density*—the more reinforcing it is. Apparently, the rat brain evolved not just to seek carbohydrate but to seek all types of calories—particularly foods that deliver the most calories per bite. Sound familiar?

Flavors and smells are a quick way for the brain to gather information about the nutritional quality of a food before it enters the digestive tract. Sclafani and other researchers have shown that certain flavors on the tongue can enhance conditioned flavor preferences. For example, rats form stronger conditioned flavor preferences if they can taste the sugar as they drink it, rather than having it infused directly into the stomach. The sugar's effects on the tongue combined with its effects in the small intestine cooperate to reinforce behavior. Other flavors work similarly. The meaty umami flavor associated with the amino acid *glutamate* (the principal component of monosodium glutamate, or MSG)[39] increases food desirability in

---

38 Some consider alcohol (ethanol) a fourth class of calorie-containing nutrient. Alcohol can also condition preferences in rodents and humans, although the mechanism may be somewhat different.

39 Nucleotides, the building blocks of DNA and RNA, are also recognized as umami by the tongue. This is why mushrooms taste meaty.

Calories don't just drive flavor preferences; they also drive preferences for the aromas, sights, sounds, and even *locations* that predict the availability of calories. It turns out that rats like to hang out in places where good things happen, and calories in the belly is definitely a good thing. This is how we learn to respond to our surroundings in a way that gets us what we want.

rats just as in humans, and glutamate also conditions flavor preferences when administered into the stomach. Sweetness indicates ripe fruit, and glutamate indicates protein-rich food like meat[40]—both of which are important sources of calories and other nutrients in the wild. Conversely, bitter flavors, odors of decay, and any food that has previously caused digestive distress are aversive.[41]

Together, this shows that animals don't seek food indiscriminately—they seek specific food properties that the brain *instinctively recognizes as valuable*—and most of these properties indicate high-calorie foods. These are presumably the nutrients that are critical for survival and reproduction in a natural environment, explaining why the rat brain responds to them by releasing dopamine and reinforcing the behaviors of seeking and eating them. The rat brain instinctively values nutrients that keep wild rats healthy and fertile, and it gradually learns how to obtain them efficiently, under the tutelage of dopamine.

CONDITIONING CULTURE

If you're starting to feel like you have a lot in common with rats, you're right. Humans share most of our innate food preferences with rats, which makes sense if you think about it: We're both omnivores that have been

---

40 Wild rats don't exactly sit down to a T-bone steak, but they do eat a variety of small animals and human food scraps that are rich in protein.

41 We can learn to like certain bitter flavors, such as the flavor of hops in beer. This is a conditioned flavor preference that develops when the flavor is paired with alcohol or another reinforcing property.

INNATE PREFERENCES	INNATE AVERSIONS
Calorie density	Bitter flavor
Fat	Odor of decay
Carbohydrate	Anything that causes digestive
Protein	distress
Sweet flavor	
Salty flavor	
Meaty flavor (umami)	

eating human food for many generations. Humans and rats are born liking sweet and disliking bitter flavors, suggesting that these are deeply wired adaptations that may have evolved prior to the divergence of our species seventy-five million years ago. In addition, people of all cultures enjoy the meaty flavor of glutamate, dislike odors of decay, and are repulsed by foods that have previously caused digestive distress. Humans also have a strong preference for salt (sodium chloride)—one inclination we don't share with rats.[42]

Our gastronomic similarity to rats is supported by the research of Leann Birch, a childhood obesity researcher at the University of Georgia, whose research team set out to see if they could extend Sclafani's results to humans.[43] Their findings, and those of other researchers, show that certain nutrients, especially fat and carbohydrate, can indeed condition flavor preferences in our own species. In other words, they reinforce behavior.

In the box above, I've listed the innate food preferences and aversions that all humans share (known or suspected). These are the food properties that shape our eating behavior by causing dopamine to spike in our brains, and thereby driving us to seek the flavors, smells, textures, appearances, and places we've learned to associate with food that contains those properties.

---

42 Rats can be conditioned to like salt by depriving them of it, but they don't spontaneously seek it like humans do. Sodium is an essential nutrient, humans lose a lot of it through sweat, and most wild environments don't supply very much of it. This may explain why salt is the only micronutrient (essential vitamin/mineral) we can taste at the concentrations normally found in food.

43 This research was conducted when Birch was at the University of Pennsylvania.

People in China and in France enjoy sugar, salt, fat, and meaty flavors. Of course, a French person may not have much appreciation for authentic Chinese food, and a Chinese person may find the odor of strong French cheese repulsive. This is because each culture has its own set of unique flavor, smell, texture, and appearance preferences that develop as the result of reinforcement learning. Just as rats aren't born preferring cherry flavor over grape flavor, we aren't born appreciating the flavors and smells unique to our culture: We develop them as conditioned preferences. These food properties become enjoyable and motivating by virtue of their repeated association with innately reinforcing properties, such as fat and carbohydrate, particularly in childhood.

When you examine this list, you'll find it apparent that the human brain is extremely preoccupied with calories. Other than salt, every innate preference is a signal that indicates a concentrated calorie source. This is presumably because calorie shortage was a key threat to our ancestors' reproductive success, so we've been wired by evolution to value high-calorie foods above all others. Consistent with this, modern-day hunter-gatherers living in a manner similar to how our distant ancestors would have lived don't spend much time or effort gathering wild brussels sprouts. They primarily spend their time looking for calorie-rich foods that can sustain their energy needs, like nuts, meats, tubers, honey, and fruit—a topic we'll return to later.

Our innate food preferences readily explain why children like ice cream and not brussels sprouts. As far as the ventral striatum is concerned, the fact that brussels sprouts are loaded with vitamins and minerals counts for approximately nothing, because they scarcely deliver any calories.[44] In contrast, we crave ice cream because our brains know that its flavor, texture, and appearance predict the delivery of a truckload of easily digested fat and sugar. Having evolved in an era of relative food scarcity, the human brain interprets this as highly desirable and draws us toward the freezer.

---

44 We can speculate about why this might be the case. There isn't any refined food in a wild environment, and hunter-gatherers tend to have a fairly diverse diet, so it was probably difficult for them to meet their energy needs without also meeting their need for all other essential nutrients. This is supported by the fact that existing and historical hunter-gatherer groups tend to have very low rates of vitamin and mineral deficiency. There may have been no need to evolve a system to seek vitamins and minerals because deficiencies weren't a major threat to reproductive success, particularly when compared to the much higher risk of insufficient calories. Salt may be an exception. We're probably wired to taste and enjoy it, like certain other animals, because it was a limiting nutrient in the wild. In contrast, we haven't evolved the ability to taste and enjoy vitamins or most minerals.

Nonconscious parts of the brain perceive certain foods as so valuable that they drive us to seek and eat them, even if we aren't hungry and even in the face of a sincere desire to eat a healthy diet and stay lean. We crave dessert even after a large meal. We crave a soda with lunch. We crave that extra slice of pizza. Willpower often bows before the power of dopamine-reinforced sensory cues, such as the dessert menu on the table, the sight of the soda machine, or the aroma of pizza. And sometimes, the reinforcement process spirals out of control.

VERY STRONG HABITS

Although reinforcement is a natural process that evolved to guide us through life successfully, it can sometimes get out of hand. As McDevitt's experiment demonstrated, when dopamine reaches very high levels in the ventral striatum, it reinforces behaviors so strongly that they can take priority over natural, constructive behaviors. This is the essence of addiction.[45] It makes sense, then, that every known addictive drug either increases dopamine levels in the ventral striatum or stimulates the same signaling pathway in a different manner. Even relatively benign habit-forming drugs, such as caffeine, seem to act on the same pathway. As a result of repeated dopamine stimulation of the ventral striatum, obtaining and using highly addictive drugs like crack cocaine can become such a high priority that it outweighs food, safety, comfort, and social relationships. The option generators responsible for seeking and smoking crack put in pathologically strong bids to the striatum, overwhelming most other bids.

According to Roy Wise, the addiction researcher we encountered earlier in this chapter, addiction is simply an exaggerated version of the same reinforcement process that occurs in everyday life. "Addiction just happens to be a very strong habit," says Wise, "because addictive drugs are very strong reinforcers."

So is it possible to become addicted to food? Food triggers dopamine release in the ventral striatum, just like drugs of abuse, and we know food

---

45 Another factor in addiction is withdrawal. People with addictions are often motivated to take a drug to end unpleasant withdrawal symptoms. However, the fact that they continue to crave the drug after withdrawal is over, and frequently relapse, suggests that addiction is fundamentally about reinforcement.

can be strongly reinforcing. Yet this idea remains extremely controversial among researchers. How is it possible to become addicted to a substance that's essential to the body? Are we addicted to water and oxygen as well? Everything else that's good in our lives, from sex to a new car to a job well done, also presumably provokes dopamine release in the ventral striatum. Are we addicted to everything?

We aren't, of course, addicted to everything. Most people have a constructive relationship with most of the good things in their lives—a relationship that can hardly be described as addiction. Yet research conducted by Ashley Gearhardt and Kelly Brownell, formerly at Yale University, suggests that some people may indeed be addicted to food. Brownell's team examined the standard diagnostic criteria for addiction to nonfood reinforcers, such as drugs, sex, and gambling.[46] They used these criteria to design a questionnaire that identifies addiction-like eating behaviors, particularly focusing on loss of control, eating certain foods despite negative consequences, and withdrawal symptoms. The questionnaire includes such items as "There have been times when I consumed foods so often or in such large quantities that I started to eat food instead of working, spending time with my family or friends, or engaging in other important activities and recreational activities I enjoy" and "I eat to the point where I feel physically ill."

Among their first sample of people, most of whom were lean, 11 percent met the criteria for food addiction. Further research revealed that people who meet food addiction criteria are more likely to have obesity and more likely to exhibit binge-eating behavior. This supports the idea that the reinforcing effect of food can lead to addiction-like behavior, overeating, and weight gain in susceptible people. However, not all people with obesity meet food addiction criteria, and not all people who meet food addiction criteria have obesity, so it only offers a partial explanation for the obesity epidemic.

To understand food addiction, we need to examine the *types* of foods that trigger addiction-like behavior. As it turns out, people don't become addicted to celery and lentils. What foods are they drawn to instead? The following quote by Gearhardt and Brownell sheds light on the question:

---

46 As defined in the gold-standard psychology reference manual, the American Psychological Association's *Diagnostic and Statistical Manual of Mental Disorders*.

> High concentrations of sugar, refined carbohydrates (bread, white rice, pasta made with white flour), fats (butter, lard, margarine), salt, and caffeine are addictive substances and the foods containing these ingredients may be consumed in a manner consistent with addictive behavior. Just like drugs of abuse, these food substances may not be addictive until they are processed, extracted, highly refined and concentrated by modern industrial processes; meanwhile, combinations of these look-like-food substances may greatly enhance their addictive qualities.

Although it's controversial to call these foods addictive, it is fair to say that they provoke dopamine release in the ventral striatum. And the more concentrated they are, the more dopamine they release. The more dopamine they release, the more they reinforce behavior, and the more they reinforce behavior, the closer they bring us to addiction.

What makes this alarming is that modern food technology has allowed us to maximize the reinforcing qualities of food, making it far more seductive than ever before in human history. We now have extremely calorie-dense, carefully engineered combinations of sugar, fat, salt, and starch that would have been inconceivable to our hunter-gatherer ancestors, who had no choice but to eat simple wild foods. Certain modern foods likely provoke a larger release of dopamine than the human brain evolved to expect, leading to destructive addiction-like behaviors in susceptible people. (Yet not everyone is susceptible—a topic we'll return to shortly.)

As Gearhardt and Brownell alluded to, this is similar to how drugs of abuse are often concentrated versions of less-addictive naturally occurring substances. For example, the leaves of the coca plant are widely chewed in South America as a mild stimulant and appetite suppressant reminiscent of caffeine. However, when we extract and concentrate the active ingredient of the coca leaf, this results in a much more addictive substance: cocaine. A secondary chemical process called *freebasing* transforms cocaine into the extremely addictive drug crack cocaine.[47] Human technology has

---

47 Freebasing involves neutralizing the charge of the cocaine molecule, making it fat soluble so it can cross fatty cell membranes much more quickly. This means the brain gets a faster, larger "hit" of cocaine, and subsequently dopamine.

allowed us to concentrate and enhance the property of the coca plant that increases dopamine release and reinforces behavior, transforming it from a useful herb into a life-destroying drug. Similarly, modern food technology has allowed us to concentrate the reinforcing "active ingredients" in food to an unprecedented degree, and addiction-like behavior in a subset of people is the predictable result.

Chocolate is a prime example of the riot of reinforcing properties that characterizes modern foods. The seeds of the cacao tree, a plant native to tropical South America, are naturally extremely calorie dense, due to their high fat content. When fermented, roasted, and ground into a paste, these seeds become chocolate: a magical substance that's solid at room temperature and melts in the mouth. To mask the naturally bitter flavor of chocolate, we add a generous dose of refined sugar, and sometimes dairy. The calorie density, fat content, carbohydrate content, and sweet taste of chocolate are a powerfully reinforcing combination, but chocolate has another trick up its sleeve that makes it the king of cravings: a habit-forming drug called theobromine. Theobromine is a mild stimulant that's moderately reinforcing, like its cousin caffeine.[48] Although theobromine on its own may not be the bee's knees, when added to a substance that's already highly reinforcing, it puts many of us over the edge. It may come as no surprise that chocolate addiction is a legitimate topic of scientific research. Even those of us who aren't "chocoholics" may experience chocolate cravings, and research indicates that chocolate is the most frequently craved food among women.

Although most of us aren't literally addicted to food, recall that addiction is simply an exaggerated version of the same reinforcement process that happens in all of us. We may not be so seduced by food that it interferes with our work or family lives, but most of us are driven to eat more calories than we should, acting against our own self-interest despite our better judgment.

Reinforcement is a nonconscious process, explaining why it isn't very intuitive, but it associates with a process that we know quite well: pleasure.

------

48 Caffeine is also commonly found in sodas and presumably increases their reinforcing value.

CONTROLLING CRAVINGS

How can we fight this instinctive force that compels us to eat too much? Drug addiction research gives us important clues. One of the best-established treatment strategies for drug addiction is simply to avoid drug-associated cues. We know from the work of Ivan Pavlov and others that sensory cues that are repeatedly associated with positive outcomes become motivational triggers. When an addicted person sees a crack pipe, smells crack smoke, or walks by the street where he usually buys crack cocaine, it triggers his motivation to smoke crack—and the urge can be overwhelming. Whether or not you're addicted to food, when you walk by a bakery and see freshly baked pastries and smell the aroma wafting into your nose, it triggers your motivation to eat pastries (or whatever food you prefer). That's simply the nature of reinforcement. Yet if you don't walk by the bakery and don't experience its sensory cues, your motivation to eat pastries will be much lower, and you won't have to fight yourself to avoid eating a fattening food.

We may have a hard time fighting our nonconscious urges to eat tasty, calorie-rich foods when they're right in front of us, but with a little bit of advance planning, we can prevail without having to exert too much of our limited willpower. The key is to control food cues in your personal environment. Ultimately, a little bit of wise planning can go a long way.

PLEASURES OF THE PALATE

Food that brings us pleasure when we eat it is described as *palatable*. Palatable food tastes good. It's a sign that the brain values a food, either as a result of instinct or reinforcement learning.

The brain presumably values certain food properties above others because they would have increased the reproductive success of our ancestors.

The most highly palatable foods tend to be dense in easily digested calories and combine multiple innately preferred food properties in highly concentrated form: ice cream, cookies, pizza, potato chips, french fries, chocolate, bacon, and many others. These are the foods that are most likely to cause cravings and a loss of control over eating, because their physical properties make them exceptionally reinforcing, motivating, and palatable. Researchers have an umbrella term for this combination of effects on the brain: *food reward*. Highly rewarding foods are those that seduce us.

Not surprisingly, research suggests that people eat larger quantities of foods they like. John de Castro, a professor of psychology at Sam Houston State University, and his research team demonstrated that people living their everyday lives eat 44 percent more calories at meals they describe as highly palatable than at meals they describe as bland. The brain perceives these foods as so valuable that it motivates us to keep eating them even if we have no particular need for energy—in fact, even if we're already drowning in energy.

What happens to food intake and adiposity when researchers dramatically restrict food reward? In 1965, the *Annals of the New York Academy of Sciences* published a very unusual study that unintentionally addressed this question. Here's the stated goal of the study:

> The study of food intake in man is fraught with difficulties which result from the enormously complex nature of human eating behavior. In man, in contrast to lower animals, the eating process involves an intricate mixture of physiologic, psychologic, cultural and esthetic considerations. People eat not only to assuage hunger, but because of the enjoyment of the meal ceremony, the pleasures of the palate and often to gratify unconscious needs that are hard to identify. Because of inherent difficulties in studying human food intake in the usual setting, we have attempted to develop a system that would minimize the variables involved and thereby improve the chances of obtaining more reliable and reproducible data.

The "system" in question was a machine that dispensed liquid food through a straw at the press of a button—7.4 milliliters per press, to be exact (see figure 15). Volunteers were given access to the machine and allowed to consume as much of the liquid diet as they wanted, but no other food.

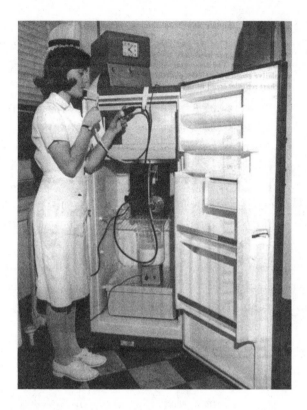

Figure 15. Food-dispensing device being tested by a nurse. Note the straw, button, and funny hat. Reproduced from Hashim et al., *Annals of the New York Academy of Sciences* 131 (1965): 654, with permission from John Wiley and Sons.

Since they were in a hospital setting, the researchers could be confident that the volunteers ate nothing else. The liquid food supplied adequate levels of all nutrients, yet it was bland, completely lacking in variety, and almost totally devoid of all normal food cues.

The researchers first fed two lean people using the machine—one for sixteen days and the other for nine. Without requiring any guidance, both lean volunteers consumed their typical calorie intake and maintained a stable weight during this period.

Next, the researchers did the same experiment with two "grossly obese" volunteers weighing approximately four hundred pounds. Again, they were asked to "obtain food from the machine whenever hungry." Over the course of the first eighteen days, the first (male) volunteer consumed a meager 275 calories per day—less than 10 percent of his usual calorie intake. The second (female) volunteer consumed a ridiculously low 144 calories per day over the course of twelve days, losing twenty-three pounds. The investigators remarked that an additional three volunteers with obesity "showed a similar inhibition of calorie intake when fed by machine."

The first volunteer continued eating bland food from the machine for a total of seventy days, losing approximately seventy pounds. After that, he was sent home with the formula and instructed to drink 400 calories of it per day, which he did for an additional 185 days, after which he had lost two hundred pounds—precisely half his body weight. The researchers remarked that "during all this time weight was steadily lost and the patient never complained of hunger." This is truly a starvation-level calorie intake, and to eat it continuously for 255 days without hunger suggests that something rather interesting was happening in this man's body. Further studies from the same group and others supported the idea that a bland liquid diet leads people to eat fewer calories and lose excess fat.[49]

This machine-feeding regimen was just about as close as one can get to a diet with zero reward value and zero variety. Although the food contained sugar, fat, and protein, it contained little odor or texture with which to associate them. In people with obesity, this diet caused an impressive spontaneous reduction of calorie intake and rapid fat loss, without hunger. Yet, strangely, lean people maintained weight on this regimen rather than becoming underweight. This suggests that people with obesity may be more sensitive to the impact of food reward on calorie intake. Is this because heightened food reward sensitivity causes obesity or because obesity increases food reward sensitivity? This question would require further research to disentangle.[50]

In 2010, Chris Voigt, the director of the Washington State Potato Commission, decided to eat nothing but potatoes and a small amount of cooking oil for sixty days. Voigt was protesting a decision by the federal Women, Infants, and Children food assistance program to remove potatoes

---

49 Although, strangely, one study reported that it works better in adults than in youths. The sample size of youths was extremely small, however (two subjects).

50 Spoiler: It's probably both.

from the list of vegetables it will pay for.[51] Voigt contended, correctly, that potatoes are actually quite nutritious—in fact, one of the few foods that provide a broad enough complement of nutrients to sustain a human in good health for months at a time.[52] He documented his journey on a Web site titled 20 Potatoes a Day, which refers to the number of potatoes he would have to eat to maintain his weight. Voigt signed himself up for two months of a starchy, bland, repetitive diet.

Despite Voigt's goal not to lose weight, the pounds melted off. He lost twenty-one pounds over the sixty-day period, much of it from his waistline. According to physical examinations before and after, his blood glucose, blood pressure, and cholesterol levels improved considerably. He had trouble eating enough to meet his energy needs because he simply wasn't hungry. While we may be tempted to question the veracity of a person whose job involves promoting potatoes, Voigt's experiment triggered an avalanche of Internet copycats who used the "potato diet" for rapid weight loss. Although anecdotal, their reports suggest that eating this bland, repetitive diet does indeed reduce spontaneous calorie intake without provoking hunger.[53] Yet this isn't just because of the blandness of an all-potato diet—there's more to the story.

THE BUFFET EFFECT

"Eat a varied diet" is a maxim that lies at the foundation of our modern approach to health. If we eat a large variety of different foods, we're likely to meet our overall nutritional needs. While this principle is sound, it also

---

51 Which, incidentally, wasn't without reason, since potatoes are usually eaten as fries and chips in the United States.

52 The main nutritional limitation of potatoes is their lack of vitamins A and B12. Night blindness due to vitamin A deficiency was common among the poor in Ireland immediately prior to the potato famine. However, the liver stores vitamins A and B12 in large quantity, so we can survive without them for a few months at a time.

53 Potatoes are one of the most quickly digested starches, meaning they raise blood sugar quickly and therefore have a high glycemic index. The effectiveness of the potato diet for fat loss is hard to reconcile with the popular belief that swings in blood glucose caused by eating quickly digested starchy foods are a major cause of hunger and overeating. Despite its pervasiveness in the scientific literature and the popular media, this idea has never had much compelling evidence to support it.

has a dark side: Food variety has a powerful influence on our calorie intake, and the more variety we encounter at a meal, the more we eat.

The effect of food variety on food intake relates to a fundamental property of the nervous system called *habituation*. Habituation is the simplest form of learning—one shared by all animals with a nervous system. It probably evolved along with the first nervous systems nearly seven hundred million years ago, since it exists in our truly ancient relatives the jellyfish.

Habituation is one of the key tricks we use to distinguish important events from unimportant noise, and it's simply this: The more we're exposed to a stimulus within a short period of time, the less we respond to it. This was demonstrated in a classic series of experiments in human infants. An infant was seated in his mother's lap, while a screen in front of him intermittently displayed a black-and-white checkerboard pattern. Researchers recorded the amount of time the infant's gaze remained fixated on the pattern each time it appeared. As any parent might have guessed, the infants paid a lot of attention to the pattern at first and then spent less time fixating on it with each exposure. When a stimulus is new, we tend to be very interested in it because it might be important. Once we've seen it many times in a short period of time, it's less likely to be important, and we stop paying attention.

As it turns out, this habituation process operates each time we sit down to a meal. In a pioneering 1981 study by Barbara Rolls and colleagues, volunteers rated the palatability of eight different foods by tasting a small amount of each, and then were provided one of the foods for lunch. After lunch, they once again rated the palatability of the same eight foods by tasting them. Rolls found that the palatability rating of the food the volunteers had eaten for lunch decreased much more than the palatability rating of the other seven foods they hadn't eaten. When the volunteers were presented with an unexpected second course containing all eight foods, they tended to eat less of the food they had just eaten for lunch. This shows that we can eat our fill of a specific food and feel totally satisfied, but that doesn't mean we won't eat other foods if they're available. Rolls called this phenomenon *sensory-specific satiety*. *Satiety* is the sensation of fullness we get after we eat food, and *sensory-specific* means this fullness only applies to foods that have similar sensory properties (sweet, salty, sour, fatty) to the ones we just ate.

BEATING THE BUFFET EFFECT

The fact that sensory-specific satiety drives us to overeat suggests a simple solution to the problem: Limit yourself to a few foods. If you find yourself at a buffet, tapas restaurant, or similar situation in which high food variety may cause you to overeat, simply choose three items you think would make a satisfying meal, and stick to them. You'll probably feel just as full on fewer calories.

Several independent researchers using various methods have confirmed that we tend to eat more total food—and gain weight—when we're presented with a large variety of foods. This goes a long way toward explaining what researchers call the *buffet effect*. We tend to overeat spectacularly at buffets, despite the fact that the food isn't always the crème de la crème.[54] At a buffet, we don't have the opportunity to habituate to any particular food, because every few bites, we're eating something new. The brain's satiety system eventually throws the emergency brake, but not before we've eaten far too much.

Sensory-specific satiety also helps explain why we're happy to eat dessert even after a large meal. We're no longer hungry for savory food at all, yet when the dessert menu appears, we suddenly grow a "second stomach." We're satiated of savory foods, but we aren't satiated of sweets. A novel sensory stimulus with an extremely high reward value makes it easy to pack away an additional 200 Calories of dessert. So it makes sense that the converse is also true, as we saw with the potato diet: When food reward and variety decrease, so does food intake.

SMOKING JOINTS FOR SCIENCE

"Numerous anecdotal accounts indicate that marijuana increases appetite and food intake in humans," begins a 1988 paper by substance abuse

---

54 Also: Thanksgiving.

researcher Richard Foltin and colleagues, referring to a phenomenon known to smokers as "the munchies." But could Foltin's team replicate this effect scientifically, or was it simply stoner lore? For thirteen days, Foltin and his colleagues confined six men to a laboratory setting where all food was provided and accurately measured. Each day, volunteers smoked either "two cigarettes containing active marijuana" or two placebo joints containing no marijuana. The primary psychoactive ingredient of marijuana, Δ^9-tetrahydrocannabinol (THC), activates the cannabinoid receptor type 1 (CB1), which plays a key role in the brain circuitry that regulates food reward. If these circuits really do have a major impact on food intake and adiposity, then activating them using marijuana should have a clear effect.

The outcome of Foltin's study was unequivocal: The men ate 40 percent more calories while they were stoned than while they were sober, and their body weights also climbed rapidly. Interestingly, they didn't overeat at meals, but instead ate more highly palatable sweet snacks, such as candy bars, between meals. A number of other studies have confirmed that marijuana increases food intake, including my favorite one: "Effects of Marihuana on the Solution of Anagrams, Memory and Appetite." I suppose even getting stoned, playing games, and pigging out is worthy of scientific study.

If THC activates the CB1 receptor and this increases food intake and adiposity, then it stands to reason that blocking the CB1 receptor should reduce food intake and cause weight loss. This is exactly the rationale for the CB1-blocking drug rimonabant, or "reverse marijuana" as I like to call it.[55] As predicted, rimonabant reduces food intake and causes weight loss in a variety of animals, including humans.

Although the drug has demonstrated its effectiveness in a research setting and it was briefly approved as a weight-loss drug in Europe, it's not currently approved for the treatment of any condition due to ongoing concerns about its negative side effects. Shockingly, "reverse marijuana" seems to increase the risk of depression, anxiety, and suicidal thoughts. Nonethe-

---

55 Rimonabant is technically an "inverse agonist" of the CB1 receptor, meaning it doesn't just keep the receptor from being activated by its natural ligands; it actually reduces the receptor's intrinsic activity.

less, marijuana and rimonabant illustrate the powerful pull the reward system exerts on our behavior, including how much we choose to eat.

Yet, if food reward causes us to overeat and we're all surrounded by it, then why do some people develop obesity while others don't?

LEAD FOOT, WORN BRAKES

A young woman stares intently at a computer screen in Leonard Epstein's lab at the University at Buffalo in New York. She's playing a slot machine computer game. Each time she clicks the mouse button, three columns of shapes spin on the screen and then settle. If the shapes are different from one another, she gets nothing—but if they line up, she gets a point. While it may sound like she's goofing off, in fact she's participating in a series of fascinating studies that are beginning to shed light on why some people become obese and others don't.

Once she earns two points, she gets a small piece of a candy bar. While she only needs two points to receive a candy bar the first time, the next time, she must earn four points to receive the same candy bar, and then eight. "We keep increasing the work requirements," explains Epstein, "until eventually the person says, 'Jeez, it's just not worth it.'" The number of responses she's willing to make before quitting quantifies how hard she's willing to work for food.[56]

To compare this to how hard she's willing to work for a nonfood reward, the researchers also give her simultaneous access to a second computer in the room. This computer is programmed with the exact same game, except it earns the woman access to an appealing magazine for a few minutes, rather than candy. The woman can switch between the two computers

---

[56] Overall food motivation reflects the sum of several different submotivations, including hunger and food reward (which, confusingly, influence one another). Epstein's experiments are designed to minimize the hunger component by giving all volunteers a snack before the experiment so that differences in food motivation are more related to food reward than to hunger. In cases where they measured and reported baseline hunger, it didn't differ between groups that exhibited different levels of food motivation, suggesting that differences in hunger probably didn't have a major impact on the results. However, it's not possible to completely exclude differences in hunger as a contributing factor.

anytime she wants, and once she decides neither reward is worth the effort, the experiment is over.

This simple technique allows Epstein's team to calculate a personal characteristic called the *relative reinforcing value of food* (RRV_{food}). RRV_{food} is a measure of how hard a person is willing to work for food, relative to a nonfood reward such as reading material—and people differ greatly in this regard. "There are huge individual differences in that some people will work really, really hard to get access to food, and other people will only work a little bit," explains Epstein. The fact that RRV_{food} measures the motivational value of food *relative* to nonfoods is important, because we often have the choice to do something other than eat. RRV_{food} asks: When faced with a choice, are you more likely to eat or to do something else?

These studies have produced very provocative results: first, that sweet foods are exceptionally motivating, especially to youths. "If you let teenagers work for sweet soda," relates Epstein, "they will work *really* hard . . . People will make *thousands* of responses for a small piece of candy."

A second provocative conclusion is that people who are overweight or obese tend to have a higher RRV_{food} than people who are lean. In particular, children who are overweight or obese are much more willing to work for highly rewarding foods like pizza or candy than lean children, even if their baseline level of hunger is the same. Consistent with their heightened food motivation, people with a high RRV_{food} eat more food both in the lab and at home. People who are overweight or obese find food more motivating than lean people do, and this leads them to eat more.

However, these studies don't tell us whether a high RRV_{food} causes people to gain weight, or whether something about the overweight state causes RRV_{food} to increase. All they tell us is that the two are associated with one another. To begin to explore the question of whether a high RRV_{food} actually causes weight gain, Epstein and other researchers went back in time.

Knowing that a substantial fraction of lean children go on to become overweight, they examined whether RRV_{food} could *predict* who would go on to gain weight and who wouldn't. Their results were remarkably consistent: RRV_{food} not only predicts weight gain in children, but in every age group they examined. In one study, adults with a high RRV_{food} gained more than five pounds over the course of a year, whereas adults with a low RRV_{food} only gained half a pound. "If you look at a group of lean people and you measure the reinforcing value of food," explains Epstein, "you can predict who will gain weight." These findings suggest that people differ in their motivation to eat food, particularly highly rewarding foods, and that this is a stable personality trait that influences each person's susceptibility to weight gain over time. This offers a partial answer to the question posed earlier: Heightened food reward sensitivity does seem to contribute to overeating and fat gain over time.

Working hard for food makes a lot of sense when it's the only way to survive. Throughout most of human history, and long before, our ancestors spent the bulk of their lives collecting, hunting, growing, and eating food—and it was often hard work. Without a powerful instinctive drive to obtain and eat food, we wouldn't have survived at a time when securing food demanded so much effort. We still carry that instinct today, but in the modern world, where food is easy to get and highly rewarding, that powerful drive can often lead us to overeat. Yet, as with most traits, people differ widely in their level of food motivation.

But that's not the end of the story. Drug abuse research suggests that a person's susceptibility to addiction depends not only on how reinforcing the drug is for him but also on his ability to control his behavior in response to a craving—in other words, his *impulsivity*. Impulsivity describes a person's ability—or lack thereof—to suppress or ignore basic urges that are beyond conscious control. It's the opposite of what we commonly call self-control. A person who finds crack cocaine highly reinforcing will crave the drug intensely after he's smoked it a few times, but if he's able to prevent himself from acting on those cravings, he won't be addicted. A second person who finds crack cocaine equally reinforcing but who is highly impulsive will readily become addicted. As Epstein puts it, "If you find something really rewarding and you have really poor impulse control, you're in a lot of trouble." Epstein coined the term *reinforcement pathology* to describe the dangerous combination of high reinforcement sensitivity and high impulsivity. He explains that it's like having a "lead foot and worn brakes." This may identify why some people are more susceptible to food addiction than others, despite the fact that we're all exposed to potentially addictive foods.

On the other hand, people who have a high RRV_{food} but who aren't impulsive (lead foot and good brakes) aren't at an increased risk of overeating or weight gain. "If you have really good self-control," explains

Epstein, "you can overcome the reward value, and you can be a foodie: someone who loves food, who's a gourmet cook, but who is lean because they can regulate the amount of food."

Does reinforcement pathology actually predict eating behavior and weight gain in the real world? While most people aren't literally addicted to food, including those who are overweight and obese, the same principles of reinforcement and impulsivity should still apply to nonaddicted people. Even if you aren't actually addicted to potato chips, you may still be drawn to eat them when you aren't hungry, and your ability to suppress that urge when they're available will influence how much of them you eat. Research from Epstein's group and others supports this concept: People who exhibit reinforcement pathology are highly susceptible to overeating, and they're also highly susceptible to weight gain.

Epstein is quick to point out that there's a third important factor in addition to RRV_{food} and impulsivity: the presence of highly rewarding food in your personal environment. "Obviously, if you have something with very low reward value, you don't need very good brakes. If you have a liver Popsicle, you don't need self-control to not eat a liver Popsicle. If you have a grilled steak and you love meat, all of a sudden you need a lot of self-control to regulate that." The deadliest combination, therefore, occurs when an impulsive person with a high food reward sensitivity lives in an environment that's bursting at the seams with highly rewarding foods.[57] And as we will soon see, the United States qualifies as such an environment.

---

57 This describes childhood in the United States. Children are naturally impulsive and react particularly strongly to basic food reward factors like sugar, starch, and fat.

4

THE UNITED STATES
OF FOOD REWARD

In the last chapter, I explained how most of us are driven to overeat by largely nonconscious brain processes that determine our food motivation, and that the brain is motivated by specific food properties like sugar, salt, and fat. Now, let's take a closer look at how those properties have changed in the US diet over the years, and how those changes help explain our surging calorie intake.

The industrial era, which has covered roughly the last two hundred years in the United States, has been the shortest period of human history. During this time, technology increased the efficiency of agriculture so dramatically that few of us have to be farmers anymore. Without industrialized agriculture, I wouldn't be writing this book, and you wouldn't be reading it. But industrialization has gone far beyond simply increasing the efficiency of agriculture—it has profoundly transformed food processing, distribution, and preparation.

Of the 2.6 million years since our genus *Homo* emerged, we were hunter-gatherers for 99.5 percent of it, subsistence-level farmers for 0.5 percent of it, and industrialized for less than 0.008 percent of it. Our current food system is less than a century old—not nearly enough time for humans to genetically adapt to the radical changes that have occurred. Our ancient brains and bodies aren't aligned with the modern world, and many researchers believe this *evolutionary mismatch* is why we suffer from such high rates of lifestyle-related disorders, such as coronary heart disease, diabetes, and obesity.

Unfortunately, we can't go back in time to observe the diets and eating habits of our distant ancestors, and today we know only a few basic facts about what and how they ate. Yet current and historical nonindustrial cultures can provide us with a rich tapestry of clues about what life may have been like for our hunter-gatherer and subsistence farmer forebears. Let's consider two examples and see what we can learn.

!Kung San

In the 1960s and 1970s, the anthropologist Richard Lee conducted a detailed study of a group of !Kung San hunter-gatherers in the Kalahari desert of Botswana, including descriptions of food collection, preparation, and eating practices. At the time, the !Kung San didn't live so differently from how our ancestors did prior to the development of agriculture, with the exception of a few tools and foods obtained through trade. They relied on a wide variety of wild animal and plant resources for food, including large and small game, insects, nuts, fruit, starchy tubers, fungi, leafy greens, and honey. Although they recognized at least 105 plant species as edible, most of their plant food intake came from only 14 species. Approximately 40 percent of their calorie intake came from meat, with liver being particularly prized. But their primary source of food was the mongongo tree, which supplied about half of the !Kung San's year-round calorie intake. The mongongo tree produces abundant quantities of a sugar-rich fruit containing a fat- and protein-rich nut. The fruit tastes similar to a date, and the flavor of the roasted nut has been described as "not unlike that of dry roasted cashews or almonds."

With few exceptions, the !Kung San processed all foods in some way prior to eating them. They used cracking to breach the extremely hard shell of the mongongo nut after roasting. They pounded foods to increase the digestibility and palatability of tough or fibrous items, including starchy tubers, mongongo nuts, and tough meats. They sometimes pulverized foods as a means of combining flavors and textures to increase palatability. For example, the !Kung San would mash together roots and

Figure 16. A !Kung San man gathering mongongo nuts. Reproduced from *The !Kung San*, by Richard Lee, with permission from Cambridge University Press.

mongongo nuts, resulting in a cheesy substance. And as with all human cultures, fire was a key food-processing method for the !Kung San, who boiled and roasted pieces of meat. Prior to acquiring pots through trade, the only available cooking method was roasting, which Lee describes as follows:

> A root or a piece of meat is carefully buried in a mixture of hot sand and glowing embers at the edge of the fireplace. The appearance of steam rising from the mound of sand indicates that cooking is proceeding. When the food is done after 5 to 30 minutes, depending on size, it is removed and knocked sharply against a rock or log to dislodge adhering sand and coals; then it is scraped to remove charred parts. Inevitably, some sand and ash are eaten with roasted food.

Appetizing, isn't it?

The !Kung San didn't have many of the culinary tools we use to dress up food today: they rarely used such added flavorings as herbs and spices, they didn't use salt, and they only occasionally had access to concentrated fat to add to their food. The !Kung San were rarely faced with serious calorie shortages; however, they were sometimes forced to rely on foods they weren't particularly enthusiastic about when they exhausted their favorite foods in a particular location. They sometimes had access to highly palatable treats, including the Kalahari truffle and honey, but due to the limitations of living in a natural environment, they ate certain other foods daily "without much enthusiasm."

When maintaining this traditional lifestyle, they were extremely lean.[58] In stark contrast to Western populations—which tend to be leanest in youth and grow wider with age—the !Kung San reached their maximum weight during peak reproductive years and declined thereafter.

Yanomamö

The Yanomamö are nonindustrial farmers who live in small permanent settlements that collectively straddle the border of Venezuela and Brazil in the Amazon basin. The anthropologist Napoleon Chagnon lived with and studied them extensively over a quarter-century period beginning in 1964. During this time, their primary foods were cultivated starches, including plantains, sweet potatoes, cassava, corn, and a variety of taro, with green and ripe plantain supplying the most calories. They also cultivated avocados, papayas, and hot peppers, but ate these in smaller quantities. They supplemented their starchy diet with a wide variety of wild animal foods, including large and small game, fish, insects, eggs, and honey. Their diet also included an assortment of wild plant foods, particularly fruit, nuts, starchy tubers, palm heart, and mushrooms.

The Yanomamö diet, as with most nonindustrial cultures, was very practical. Chagnon described their attitude toward food preparation as follows: "In general, the Yanomamö prefer foods that require little processing—a kind of 'take it from the vine and throw it on the fire' attitude that applies to vegetable and animal food alike." They used four food preparation methods—roasting, boiling, smoking, and grating—but they didn't use

---

58 Mean BMI was below 20 kg/m^2 in both genders and at all ages.

added flavors, fats, or salt, and they did little beyond cooking to increase the palatability of their foods.[59]

Because of their low salt intake, the Yanomamö were included in the INTERSALT study, an international study of salt intake and blood pressure, which determined that Yanomamö blood pressure remains remarkably low throughout life. There's no way to know how much of this is due to their low salt intake as opposed to other factors, such as their high level of physical activity. However, lifelong low blood pressure is typical in nonindustrial cultures. The Yanomamö were relatively lean,[60] and neither Chagnon nor other researchers reported any obesity among them, despite the fact that they tended to have adequate food supplies.

WHAT DO NONINDUSTRIAL DIETS HAVE IN COMMON?

As I hope our brief tour of the eating habits of two nonindustrial cultures has illustrated, the diets of our distant ancestors were probably radically different from ours today. The diets of nonindustrial cultures vary widely, yet they share important commonalities that fundamentally differentiate them from the diets of people living in modern affluent nations. If we can identify these commonalities, we may be able to understand what the diets of our ancestors were like, and in turn, what our bodies and brains are adapted to. Here are three prominent characteristics these diets have in common:

First, they include a limited variety of foods. For example, although the !Kung San recognized at least 105 plants as edible, only 14 formed the bulk of their plant food intake, and only a subset of these 14 plant foods was available at any given season and location. Throughout the year, half of their calorie intake came from a single food, the mongongo fruit/nut. Over the course of the entire year, the !Kung San diet was quite varied; yet over the course of a day, it may have focused on only a few foods.

---

59 Despite the fact that they grew hot peppers, Chagnon doesn't describe the Yanomamö using them as a spice, so it must not have happened frequently.

60 Average adult BMI of 21.5 kg/m^2 in men and 20.8 kg/m^2 in women.

Because of the seasonal availability of key resources, the same is true of most other nonindustrial cultures.

Second, they have a limited ability to concentrate the reinforcing properties of food. With only the most basic processing methods at their disposal, nonindustrial cultures—and presumably our distant ancestors—are forced by necessity to eat food in a less calorie-dense, less refined, less rewarding state. Most don't have the ability to add refined starch, sugar, salt, or concentrated fat to their meals. The glutamate they eat comes from cooking meat and bones rather than from crystalline MSG. Added flavors, such as herbs and spices, are limited. Although we can find isolated examples of traditional cultures that use concentrated fats, salt, multiple spices, sugars, or more refined starches, none boasts all the enhancements of the affluent industrial diet.

Third, they use few cooking methods. The cooking methods of nonindustrial cultures are extremely limited by modern standards, with most cultures only using two or three methods. Even in affluent Western cultures, cooking methods were limited by technology until relatively recently. Until the 1820s, most cooking in the United States took place in an open hearth, which is a time- and labor-intensive method that makes complex cooking techniques difficult. Cast-iron wood- or coal-fired stoves replaced open hearth cooking in the 1820s, and they remained the dominant cooking method until they were replaced by gas and electric stoves in the 1920s. Techniques as simple as sautéing and temperature-controlled baking were difficult or impossible to perform in the home prior to these advances in technology.

To the modern palate accustomed to constant entertainment, nonindustrial diets, and likely the diets of our distant ancestors, would seem repetitive, bland, and sometimes unpalatable. Fortunately for our palates but not for our adiposity, we live in different times.

FOOD REWARD IN THE UNITED STATES

If we're to believe that food reward has played a role in overeating and our expanding waistlines, we must have evidence that the food properties that titillate our brains' reward circuits, and/or the food cues that make us seek them, have increased over time. Unfortunately for the American waistline, this evidence isn't hard to come by.

A hunter-gatherer walking through a modern grocery store would be bewildered by the dizzying array of food choices, particularly those of high calorie density and palatability (not to mention all those boxes adorned with strange cartoon characters). The Food Marketing Institute reports that in 2013, the average US grocery store contained a staggering 44,000 items, up from an already impressive 15,000 items in 1980. In sharp contrast to the dietary habits of nonindustrial cultures, which are limited by availability, affluent industrial cultures are steeped in a vast abundance of food choices, most of which are professionally crafted to maximize reward value. This variety means we experience less sensory-specific satiety, almost like a perpetual buffet.

The magnitude of the change in US food habits over time, as demonstrated by USDA food-tracking data, is difficult to overstate. One of the things the USDA keeps tabs on is the proportion of food spending we dedicate to food consumed in the home versus outside the home, and this provides us with a rough approximation of how much people are cooking at home versus going out to eat. In 1889, Americans spent 93 percent of their food expenditures on food to be eaten at home, and only 7 percent eating out. Today, we spend about half of our food expenditures on food to be eaten at home, and the other half eating out (see figure 17). Much of the recent increase has come from fast-food spending, which has increased ninefold since 1960. These figures actually underestimate the magnitude of the change in US food culture, because today many of the foods we eat at home are actually commercially prepared, such as pizza, soda, cookies, and breakfast cereal.

It's clear that our food culture has changed profoundly in this country over the last century, a period over which we've outsourced the majority of our food preparation to professionals. This shift toward outsourced food preparation on an industrial scale has led to remarkable changes in the processing and composition of the food we eat.

So how exactly did we get hooked on processed food? Let's return to what we previously discussed about food reward. We know that the brain

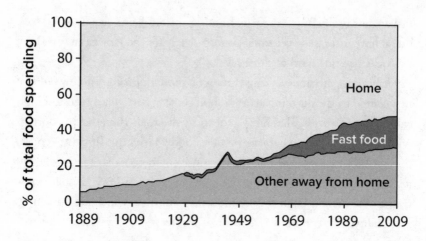

Figure 17. Percentage of food expenditures dedicated to food consumed at home, away from home, and at fast-food restaurants in the US between 1889 and 2009. The "other away from home" category includes all expenditures that were neither fast food nor food eaten at home, such as sit-down restaurants. Data are from the USDA Economic Research Service.

recognizes calorie density, fat, carbohydrate, protein, sweet flavors, salty flavors, and meaty flavors as innately reinforcing and that those properties drive our motivations, preferences, and habits. Examining the trajectory of food processing over human history, it's clear that we've gradually purified these reinforcing attributes to their most concentrated states to satisfy our own palates. In nonindustrial diets, fat, starch, sugar, salt, and free glutamate rarely exist in highly concentrated form. In modern industrial diets, they're used in nearly pure form as ingredients—and married together in irresistible combinations. To illustrate this, let's take a closer look at sugar, fat, and glutamate.

Sugar holds a special place in the brain's reward centers, perhaps because in the time of our distant ancestors, sweetness signified fruit or honey—both safe and valuable sources of nutrition. These were the only two sources of sweetness for most of human history. But we gradually figured out how to extract pure sugar from beets and sugarcane, although initially it was so expensive that only the wealthy ate it regularly. As technology advanced, concentrated sugar became cheaper, easier to obtain, and easier to use. In the 1870s, granulated sugar became widely available in the United States, making it more convenient to add sugar to food. A glassblowing machine patented in 1899 led to the mass production of bottles, reducing the cost of sweetened beverages. Then, in the 1920s, refrigerated vending machines were invented, making it easier than ever to grab a cold, refreshing soda. The 1970s brought one of the most significant technological advances in US diet history: high-fructose corn syrup. This is a sweetener produced from corn starch that has approximately the same sweetness as cane sugar. Thanks to government-subsidized corn, it's so cheap that food manufacturers can use it to beef up the reward value of their foods at virtually no cost, tickling the brain circuits that make us reach for the cookies.

My research partner Jeremy Landen and I stitched together records from the USDA and the US Department of Commerce to form a complete picture of added sweetener intake in the United States from 1822 to 2005 (see figure 18). These data don't include the natural sugars that come from fruit and vegetables, but they do include honey, cane sugar, beet sugar, and high-fructose corn syrup. Looking at this graph, it's clear that we eat far more added sugar today than we did in the early 1800s. In 1822, we consumed the amount of added sugar in one twelve-ounce can of cola every five days, whereas today we consume that amount every seven hours.[61] Again, advances in food technology and the increasing influence of the food industry are behind this shift. The competitive nature of the food market dictates a race to the bottom in which manufacturers converge on the most rewarding concentration of added sugar in each food: not too little and not too much. This optimal concentration is called the sugar "bliss point," and it's the subject of much industry research, as detailed in the excellent books *Salt Sugar Fat* and *The End of Overeating,* by Michael Moss and David Kessler, respectively.

As discussed in the last chapter, added fats are another highly effective way to rev up the brain's reward circuits. Isolated fats, such as soybean oil, canola oil, and butter, increase the calorie density and reward value of foods at little additional cost, which is why they're used liberally in commercially prepared foods, including restaurant food. USDA data reveal that total fat intake has increased modestly in the United States over the

---

61 This includes the time we're asleep. Stated differently, today we consume the amount of added sugar found in 3.4 cans of cola every twenty-four hours, on average. Some people consume much more, others much less.

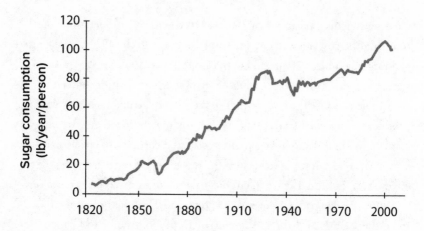

Figure 18. US added sweetener consumption per person, 1822–2005. Sweetener consumption is represented in pounds per year, per person. Data are adjusted for waste at a rate of 28.8 percent. Data are from the USDA Economic Research Service and the US Department of Commerce. Special thanks to Jeremy Landen.

course of the last century, but *added* fat intake has doubled (see figure 19). The types of fats we use in our cooking have also changed profoundly—animal fats, such as butter and lard, have largely been supplanted by refined seed oils (vegetable oils) such as soybean oil. Rather than getting our fat from whole foods like meat, dairy, and nuts, we now get it primarily from oils that are mechanically and chemically extracted from seeds. These liquid oils are cheap and convenient to add to foods that would otherwise contain little fat, creating such food reward masterpieces as french fries and Doritos. Added fat increases the calorie content of food and our drive to eat it, ultimately causing us to overeat.

Free glutamate, responsible for the meaty umami flavor, occurs naturally—in small amounts—in cooked meats and bone stock. Today, the food industry uses highly concentrated crystalline MSG to give food the savory, meaty flavor our brains crave. To get around the health concerns that surround MSG, companies have developed alternative sources of glutamate that slip under our radar, such as hydrolyzed yeast and soy protein extracts. The purpose of these substitutes is the same: satisfy the brain's innate preference for glutamate that keeps people coming back to flavored tortilla chips, salad dressing, soups, and many other foods.

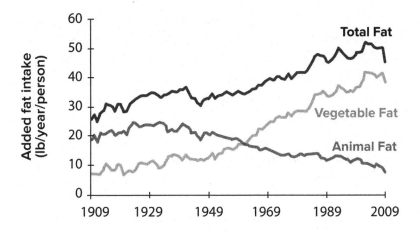

Figure 19. US added fat intake per person, 1909–2009. The animal fat category includes butter, margarine, lard, and tallow. Tallow data are not available before 1979, but intake at that time was low. The vegetable fat category includes salad and cooking oils, shortening, and "other edible fats and oils." Data are from the USDA Economic Research Service and are adjusted for waste at a rate of 28.8 percent, and for an artifact in 2000.

IDENTIFYING THE ENEMY: TOO MUCH SUGAR OR TOO MUCH FAT?

In popular media, there is a perennial debate over whether sugar or fat is responsible for the obesity epidemic. This has led some people to view obesity research as a team sport rather than a scientific discipline. Allow me to end the debate by stating what most researchers find quite obvious: It's both. In particular, the combination of concentrated sugar and fat in the same food is a deadly one for our food reward system. It's also a pairing that rarely occurs in nature, so it's tempting to speculate that it's more than our brains are equipped to handle constructively. Think ice cream, chocolate, cookies, and cake—foods we crave, and foods we don't need to be hungry to eat. What would they be without sugar, and what would they be without fat?

Although MSG has a bad reputation among health-conscious consumers, humans have always sought glutamate, and throughout history we've found ways to get it.[62] The first source of glutamate, perhaps hundreds of thousands of years old, was simply cooked meat. Once we invented cooking vessels, we began boiling bones to make a meaty, glutamate-containing broth—something we've probably done since before recorded history. The next step was to develop fish sauce, which is very high in glutamate as a result of the decomposition of naturally occurring fish proteins. Primarily associated with traditional Asian cuisine today, the ancient Romans used a similar sauce called *garum* more than two thousand years ago. Soy sauce, another concentrated source of glutamate, has been popular in parts of Asia since the same era. The isolation of pure glutamate by Tokyo Imperial University researcher Kikunae Ikeda in 1908 and the subsequent commercial production of MSG were the culmination of a long historical process that gave us access to increasingly concentrated forms of glutamate. As with other reinforcing food properties, technology allowed us to distill the umami flavor down to its primary active ingredient, yielding a powerful tool for increasing the reward value of food.

But the march of progress hasn't stopped there. Modern food chemists have created or isolated a vast number of seductive flavors that compel us to buy food. Although these usually appear on an ingredient label as the cryptic phrase "artificial flavors" or "natural flavors," such terms often conceal a carefully engineered suite of dozens of ingredients.

Habit-forming drugs, which are reinforcing due to their direct actions on the brain's dopamine system,[63] are also more prevalent in modern affluent foods. Caffeine and alcohol are inherently reinforcing, meaning that they drive conditioned food and beverage preferences that can contribute to overeating. Alcoholic beverages are rich in calories—ranging from 90 to

---

62 The evidence that MSG is directly harmful to health isn't particularly compelling. People commonly attribute uncomfortable post-meal symptoms to MSG, but double-blind trials have generally failed to support this idea. Much of the hoopla over MSG relates to its ability to cause damage to a brain region called the hypothalamus, and therefore cause obesity, when injected into newborn animals (via a phenomenon called excitotoxicity). However, this appears to be irrelevant to typical levels of MSG in the diet, which don't elevate blood levels, much less brain levels, of glutamate significantly.

63 Or closely related pathways such as the adenosine receptor.

180 Calories per serving. To put that into context, two beers per day can be the difference in calorie intake between a lean person and an overweight person.[64] Most American adults drink alcohol regularly, and we virtually never drink it to satisfy a need for calories. We drink it because we like it, whether we're hungry or not. Caffeine itself doesn't contain calories, but it's often associated with calorie-rich cream and sugar—and as with alcohol, we don't take in these extra calories because we're hungry for them. And theobromine, the mild habit-forming drug in chocolate, also drives us to eat excess calories.

Finally, cooking techniques have proliferated since the time of our distant ancestors, increasing the tools available to maximize food reward. Whereas most nonindustrial cultures have two or three simple cooking methods at their disposal, chefs today can bake, roast, broil, sauté, sear, deep-fry, poach, steam, boil, sous vide, pressure-cook, slow-cook, flambé, smoke, or grill—and pair these techniques with a wide variety of ingredients, including added fats like butter, sweeteners like sugar, and flavorings like salt.

It's no accident that our modern food landscape looks this way: The innate preferences of the human brain have dictated it via consumer demand. The food industry competes fiercely for your "stomach share" by pushing your food reward buttons, trying to turn you into a repeat customer. It does this largely by appealing to the things your brain intuitively wants. Highly reinforcing food, and the use of food cues in advertising that trigger us to purchase it, are the inevitable outcome of a competitive economy in which companies jockey for our food spending.

Modern food technology gives us an exquisite degree of control over the rewarding properties of food, and it offers us tremendous food variety. As food technology has advanced, affluent diets have gradually evolved to mirror the innate food preferences of the human brain. While these preferences kept us alive and well in an ancestral environment, they

---

64 The difference in calorie intake between a lean and an overweight person of the same gender and height is about 10 percent. Assuming two beers provide 300 Calories, and the average lean person requires about 2,400 Calories to maintain weight, two beers is more than enough to bridge the calorie gap between lean and overweight.

cause us to overeat in an environment that satisfies them to an unprecedented degree.

SUPERNORMAL SEDUCTION

Technology has allowed us to create foods that are far more seductive than those that occur in a natural environment, and this leads to an equally unnatural response from the brain. The idea that our hardwired inclinations can be overstimulated by unnaturally powerful cues goes back to the 1930s. During a study of the nesting behaviors of the ringed plover, Koehler and Zagarus noted that the birds prefer to sit on artificial eggs that are an exaggerated version of natural eggs. A typical ringed plover egg is light brown with dark brown spots. When Koehler and Zagarus offered the birds artificial eggs that had a white background and larger, darker spots, they readily abandoned their own eggs in favor of the decoy. Similar experiments with oystercatchers and herring gulls showed that they prefer larger eggs, to the point where they will leave their own eggs in favor of absurdly large artificial eggs that exceed the size of their own bodies.

Essentially, birds have an innate preference for specific egg characteristics. Ringed plovers are wired to prefer a roundish egg with contrasting spots, such that a roundish egg with larger spots and more contrast elicits a stronger preference. Oystercatchers and herring gulls are wired to prefer large eggs, possibly because larger eggs tend to be healthier, and their preference for larger eggs extends well beyond the natural range. The Dutch biologist Nikolaas Tinbergen coined the term *supernormal stimulus* to describe the phenomenon whereby, as he put it, "it is sometimes possible to offer stimulus situations that are even more effective than the natural situation." Whatever a species' innate preferences are, they can often be overstimulated by presenting a cue that's more powerful than what the species has evolved to expect—and this can sometimes lead to highly destructive behavior. It seems likely that certain human innovations, such as pornography, gambling, video games, and junk foods, are supernormal stimuli for the human brain.

In nature, supernormal stimuli are sometimes used as tools for exploitation. For example, the common cuckoo is a parasitic bird that exploits the innate preferences of its host species, one of which is the reed warbler.[65] The egg of a cuckoo looks almost identical to a reed warbler egg, except it's larger. Once hatched, the larger cuckoo chick ejects all reed warbler eggs and chicks from the nest. It begs for food using a song that closely resembles the sound of a whole brood of reed warbler chicks, tapping into its host's hardwired feeding response to the sound of multiple hungry chicks. The large size and bright color of the chick's mouth as it begs may also stimulate the parent's desire to feed it more than a normal chick. By the time the cuckoo is ready to fledge, it's quite a bit larger than its adoptive parent, which it has profoundly exploited for its own gain.

Likewise, our own innate food preferences are commercially exploited by concentrating and combining the properties we find most rewarding, resulting in foods that are more seductive than what our ancestors would have encountered. Creating an obesity epidemic wasn't the objective; it was just an unfortunate side effect of the race to make money.

COLLATERAL DAMAGE

The tremendous influence of food reward on our eating habits becomes obvious when we take a close look at the American diet. The 2010 USDA *Dietary Guidelines for Americans* reported that the following six foods are the top calorie sources for US adults, in descending order of the number of calories that each contributes to our diet:

1. Grain-based desserts
2. Yeast breads
3. Chicken and chicken mixed dishes
4. Soda/energy/sports drinks
5. Alcoholic beverages
6. Pizza

---

65 The cuckoo also parasitizes other species of birds. A cuckoo's egg tends to match the pattern of the species it regularly parasitizes.

Topping the list are "grain-based desserts," a category that includes cake, doughnuts, and cookies. Next is bread, which is often viewed as an innocuous food but is usually made of refined white flour and is surprisingly calorie dense.[66] Bread also happens to be the single largest source of salt in the US diet. Then comes "chicken and chicken mixed dishes," which may owe its prominent position on the list to our fondness for fried chicken and chicken nuggets.[67] Fourth on the list is soda/energy/sports drinks, or as I like to call it, sugar water. Fifth is alcoholic beverages, and sixth is pizza. Where are the fruit? Beans? Nuts? Most of the foods on this list are calorie-dense combinations of refined sugar, refined starch, concentrated fat, salt, and habit-forming drugs (caffeine and alcohol).

Because of how our brains are wired, we don't form celery habits. We form cookie habits and pizza habits. These are the foods that make us want to eat more of them, eventually establishing deeply ingrained eating patterns that are hard to change. We don't eat these foods for their health benefits—we eat them because they reinforce our behavior.

Here are the top six sources of calories for children and adolescents:

1. Grain-based desserts
2. Pizza
3. Soda/energy/sports drinks
4. Yeast breads
5. Chicken and chicken mixed dishes
6. Pasta and pasta dishes

This list is similar, but perhaps even more heavily weighted toward highly rewarding foods at the top. These are the foods our children are eating. In light of this, is it so surprising or mysterious that many of them have obesity?

Yet the reach of the food industry doesn't stop in the dining room: It also extends into our living rooms, roadways, and offices.

---

66 This is counterintuitive, because most people think of bread as a light, fluffy food. In reality, as soon as you chew it, the air goes away, and it's quite calorie dense.

67 Nutrition researcher Marion Nestle offered this speculation in her book *Why Calories Count*.

MANUFACTURING CRAVINGS

As we discussed, once you've eaten a rewarding food a few times, cues associated with that food can trigger your motivation to seek it. For example, when we smell french fries, see images of them, or enter a place where we previously ate them, we crave french fries. The brain is saying, "This is a situation in which you can get a valuable food," and it stimulates your motivation accordingly. Food advertising taps into this fundamental principle, exposing us to food cues that trigger our desire to purchase and eat food—and it's highly effective.

The food industry pours a staggering amount of money into food advertising each year. In 2012, the ten largest food and beverage manufacturers alone spent $6.9 billion on food advertising, and fast-food restaurants spent more than $4 billion on top of that. To put that into perspective, the total amount of funding dedicated to obesity research by the National Institutes of Health, the primary funding agency for biomedical research in the United States, was less than $1 billion in 2012. The amount of money and effort put into convincing us to eat far outweighs that put into preventing overeating and its consequences.

The food industry spends this money because it gets results. People who view food advertisements tend to prefer, purchase, and request the advertised products. The average US adult views 20 food advertisements per day on television alone—adding up to more than 7,000 food cue exposures per year. Children are particularly susceptible to food advertisements due to their naturally high impulsivity and inability to understand persuasive intent. Children in the US view more than 12 food advertisements per day on television, totaling more than 4,300 per year. And food ads don't compel us to buy brussels sprouts! Unprocessed, low-calorie foods like vegetables are neither very profitable to sell nor particularly compelling to consumers. Food cues are most effective when they predict highly rewarding foods rich in fat, sugar, starch, salt, and other reinforcing properties, and these are the foods that appear most often in advertisements.[68]

---

68 The principles of advertising aren't limited to food reinforcement. Companies use all available tools to get us to buy their products. These can include both rational and emotional appeals, such as

From the perspective of food reward, it's difficult to imagine a more fattening food environment than the modern United States, and the rest of the affluent world isn't far behind. We're surrounded by a staggering variety of foods that are designed to appeal irresistibly to our innate food preferences, and we're continually bombarded by cues that remind us of these foods. Home cooks and the food industry both play a role in this; however, the food industry is particularly effective at it, and the commercialization of food preparation has paralleled our increasing tendency to overeat and gain weight. Our expanding waistlines are an unintended but predictable consequence of the march of progress in a competitive economy.

Yet food reward isn't the only thing that drives us to overeat. It goes hand in hand with convenience—another factor that titillates the brain circuits that determine our food intake, and one that we'll explore further in the next chapter.

---

highlighting the value or quality of a product, associating a product with positive emotional states, and appealing to social status. The latter strategy is surprisingly relevant to children, who are highly responsive to the "cool factor" of foods. This is another subject Michael Moss discusses in his book *Salt Sugar Fat*.

5

THE ECONOMICS OF EATING

As dawn breaks over the hills of Sipunga in northern Tanzania, light filters through the branches of a baobab tree. Maduru and his wife, Esta, emerge from their grass hut, joining others by the fire who have already risen. Esta sits down and nurses their one-year-old daughter. Wearing only a pair of dusty khaki shorts, Maduru sharpens an arrow while talking with other men in the camp to finalize his plans for the day.

Maduru and Esta are Hadza, a hunter-gatherer culture in the eastern Rift Valley of Africa—an area called "the cradle of mankind." The Rift Valley earned this name due to its incredible diversity of primate fossils, which trace a lineage from modern humans back to the oldest members of our genus *Homo* 2.6 million years ago. The Hadza practice a foraging lifestyle that can give us insight into how all humans lived prior to the advent of agriculture approximately twelve thousand years ago. Although the Hadza are not carbon copies of our Stone Age ancestors, they may be the closest living approximation we have. Their way of life helps us understand the challenges our ancestors may have faced, and the physical and mental adaptations they evolved to meet those challenges. In this instance, they're going to help us understand the cost-benefit calculations that drive us to overeat in the modern affluent world.

Maduru's friend Oya mentions that he came across the tracks of several kudu, a type of large antelope, while he was returning to camp yesterday. Energized by his friend's tip, Maduru collects his tools and prepares for the day's hunt. He slips on his sandals, places a small ax over his shoulder, ties a small plastic bucket onto his back, slides a knife into his belt, and grasps a bow, arrows, and fire drill in his hands. He sets off alone in the general direction of the kudu tracks.

As Maduru heads off into the hills, Esta is engaged in a lively debate with five other women in camp over where to dig tubers. Eventually, they select a location and head out together, joined by a young girl and boy. The women's technology is extremely simple: Each woman carries a sharpened digging stick approximately three feet long, a knife for keeping it sharp, and a fabric sling on her back. One woman carries embers from her hearth to use for cooking in the field. Esta carries her infant girl in her sling.

After walking about a mile, Maduru comes across the kudu tracks. They look more than a day old to him, but he follows them anyway to see if he can find fresher signs. As he follows the tracks, he is constantly making side trips to investigate whether other valuable foods may be nearby. He stops for a moment, eats a handful of sweet *undushipi* berries, and continues on his way. As he walks alongside the kudu tracks, he hears a pebble fall to his left. He spots a dik-dik, a diminutive African antelope, standing on a rocky crag a hundred feet away. It hasn't detected him yet. He crouches and slowly circles downwind, concealing himself behind plants, quietly stalking it until he's within range. He nocks his *kasama,* an arrow with a sharp, laurel-leaf-shaped metal point, quietly draws his bow, and lets the arrow fly, striking the dik-dik in the heart and lungs and killing it on the spot.[69] He collects his prize, settles under the shade of a tree, cooks and eats the dik-dik's liver, parts of its head and neck, and one front leg, and then takes a midday nap to escape the heat.

Meanwhile, Esta and her party have walked two miles to a promising location for digging tubers. Once they've arrived, they scan the rocky landscape for specific types of vines growing up a bush or tree, which signal

---

69 Hadza boys begin practicing archery at about three years old, and by age five or six, they are proficient archers. Archery skill peaks in the midthirties but remains high throughout life. A typical Hadza bow has a draw weight of about seventy pounds, with some reaching ninety-five pounds. Typical draw weights for modern recurve bows and longbows are thirty to sixty pounds.

the presence of an underground tuber. Esta sees an *//ekwa* vine climbing a large shrub, and she approaches to investigate. She examines the vine closely and then thumps the ground with the blunt end of her digging stick, listening carefully to determine the size of the tuber buried beneath. Satisfied that the tuber is worth her while, she begins digging with the sharp end of her digging stick. She digs rhythmically for ten minutes, finally pulling out a tuber that looks a bit like a long, crooked sweet potato. Each woman in the party will harvest a number of tubers of two different species. The women then use the embers they brought to build a fire and roast some of them. After cooking, they peel them and cut them into pieces. These tubers are extremely fibrous, so they chew the pieces thoroughly to extract the carbohydrate-rich juice, and spit out the fibrous quid that remains. After eating, Esta and her party take a nap in the shade of a large bush.

Feeling refreshed after his nap, Maduru continues stalking the kudu. However, after walking only a few hundred yards, he hears a buzzing sound dart by his ear: a honeybee. After some investigation, he spots a hole in a baobab tree that may contain the bee's nest. He stares at the hole until the faint glint of a single departing bee confirms that he has indeed found the nest. He uses his fire drill to make a small fire and cuts six wooden pegs from a nearby tree. He takes a smoking torch from the fire and begins hammering the pegs into the bark of the baobab tree with the back of his ax. One by one, he hammers in the pegs and uses them to climb the tree until he has reached the level of the nest. He uses the smoking torch to sedate the angry bees, enlarges the entrance to the nest with his ax, and reaches his entire arm into the hole. Although he gets stung a few times, he manages to extract a large portion of honey-filled comb. After descending the tree, he eats a pint of honey on the spot and places the rest into the bucket on his back. It's midafternoon, and he thinks the kudu are probably long gone by now, so he heads back to camp.

Upon awakening, Esta and her party place the remaining tubers in their slings and head back to camp. On the way home, they make a short detour to a baobab tree and collect fallen fruit, adding it to their slings.

In late afternoon, the men and women gradually trickle back into camp. As dusk falls, they make a fire and prepare the day's harvest, sharing with their families and neighbors as they sit around their family hearths to eat, nurse, play, laugh, and tell stories about the day. Maduru butchers

Figure 20. A Hadza man climbing a tree to reach a bee's nest. Note the wooden pegs hammered into the tree, upon which he is standing. Photo generously provided by Brian Wood.

and roasts the rest of the dik-dik. Nearly every part of the animal will be eaten, including most internal organs, and a broth made from the bones. Maduru also shares his honey, which the others drink eagerly. The women roast and peel more //ekwa tubers and distribute the pieces, and open baobab fruit by striking them against a nearby rock. They extract and eat the white, tangy, chalky baobab pulp and spit out the seeds.

This fictional account illustrates what a typical day might look like for a Hadza couple, based on the detailed work of anthropologists Frank Marlowe, Brian Wood, and others. Although it may seem like a simple story, it in fact illustrates some of the complex economic choices that face the Hadza each day—choices that have shaped the structure and function of the human brain over millions of years.

Figure 21. Hadza women processing baobab fruit. Photo generously provided by Brian Wood.

OPTIMAL FORAGING

"Life is a game of turning energy into kids," says Herman Pontzer, an associate professor of anthropology at the City University of New York who studies energy expenditure in the Hadza. Since obtaining calories is such a key requirement for survival and reproduction, it's one of the most important drivers of the natural selection process that shaped today's species. In a natural environment, there are more and less effective ways to obtain food, and animals that are more effective at obtaining food are more likely to pass along their genes to the next generation. In this way, natural selection gradually crafts brains that generate efficient food-seeking behaviors.[70]

As it turns out, biologists and anthropologists have been able to mathematically model the basic principles of efficient food seeking using a

---

70 In humans and certain other animals, inherited cultural knowledge can complement inherited genes.

discipline called *optimal foraging theory* (OFT). OFT assumes that animals have been crafted by natural selection to acquire food from their environment efficiently, and researchers have successfully applied it to a variety of different species, including human hunter-gatherers. Given the bewildering complexity of human behavior, the basic math of OFT is disarmingly simple:

$$\text{Value of a food item} = \frac{\text{calories gained} - \text{calories expended}}{\text{time}}$$

The value of a food item, and hence whether or not it's worth pursuing, depends on the number of calories it contains, minus the number of calories required to obtain and process it, divided by the amount of time required to obtain and process it.[71] In other words, a food's value is roughly determined by its calorie return rate. This is the same basic equation economists use to maximize profits, because the principles that govern efficient profit seeking are also those that govern efficient food seeking. Even though a rat doesn't understand economic theory, natural selection has crafted its brain so that it behaves *as if* it understands economic theory. Natural selection has turned rats, and humans, into unwitting economists.

The work of many anthropologists demonstrates that OFT, while imperfect, is surprisingly good at predicting the foraging behavior of human hunter-gatherers. For example, OFT predicts, and observations confirm, that hunter-gatherers rarely bother collecting foods that are low in calories. One surprising implication is that hunter-gatherers don't often collect or eat vegetables—that is, low-calorie plant foods such as leaves.[72] If you're a hunter-gatherer, it doesn't make a lot of sense to burn 200 Calories collecting 50 Calories' worth of salad.

If we examine the account of Hadza foraging above, we can identify many instances in which Maduru and Esta made important economic

---

71 From this fundamental premise of OFT, anthropologists have derived several other equations that describe specific aspects of hunter-gatherer behavior, including what food sources a culture will exploit among many possible options, how many different food sources a culture will exploit at once, and how long a group will stay in a particular spot before their returns diminish enough to justify a move.

72 There are reports of !Kung San and Hadza eating leaves; however, these are not a major part of the diet.

choices about food. Maduru set out to pursue a kudu, which is a large game animal with one of the highest energy return rates. It made sense for him to begin the search with a kudu in mind and spend a long time pursuing it, because the benefit of killing one is so great. Yet the optimal foraging strategy also involves being alert to other opportunities that may arise. For example, once Maduru spotted the smaller dik-dik, there was a good chance he could kill it and enjoy a good energy return from a short time investment. With the information that there is an unsuspecting dik-dik nearby, the energy return rate of hunting it is suddenly higher than continuing to search for the kudu. Yet if he had set out to find a dik-dik without any indication that one was nearby, he probably would have searched all day unsuccessfully. Similarly, the sound cue of the bee alerted him to a nearby nest and sent his foraging trip on a productive tangent. Honey has one of the highest energy return rates, and the Hadza will almost always drop everything to harvest it when they locate a good nest.

Esta also made important economic choices. In the morning, she consulted with the other women in her group to determine which location was most likely to yield the best energy return rate for tubers. That involved balancing the number of tubers they thought would be present at each location with the distance they would have to walk and the amount of effort they would have to exert to dig them up. When they were returning to camp, the sight of baobab fruit on the ground offered them the opportunity to collect a substantial amount of energy with little effort or time investment.

Whether or not they were fully aware of it, Maduru and Esta were carefully maximizing their calorie return rates. If this all sounds like common sense, that's because our brains are wired to understand basic economic principles.

Of course, the OFT equation doesn't explain everything. Human behavior is the result of many interacting motivations, so it's not surprising that there are exceptions to the predictions of such a simple formula. "People have goals besides calories," explains Bruce Winterhalder, professor of anthropology at the University of California–Davis. "Sometimes they're out looking for the most luxurious bright red feather that is required for a ritual." One example of this is that a food's *quality,* in addition to its quantity, can impact its value. In particular, hunter-gatherers (and nearly all populations globally) typically value meat calories more than they value plant calories.[73] Kim Hill, an anthropologist at Arizona State University whose work with Aché hunter-gatherers in Paraguay has been instrumental in adapting OFT to humans, has found that factoring this distinction into his calculations improves the ability of the OFT equation to predict foraging behavior. So even though calories are the primary consideration, the value of a food to a hunter-gatherer is determined by more than simply its calorie content.

Other factors can also come into play, such as risk. Even if the most efficient way to get calories is to pick fruit from a tall tree with weak branches, it may not be worth the risk of falling. Cultural influences such as taboos also impact food selection. For example, according to Wood, "No Hadza would pursue even a large, fatty savannah monitor lizard, because snakes and lizards are simply not considered food in Hadza culture." Nor do they eat fish because they are "like snakes." Personal preferences, hunger, and time delays can also play a role. The bottom line, according to Hill, is that "the brain is designed to take into account more than just energy acquisition." Yet given all the possible motivations the basic OFT equation ignores, it does a surprisingly good job of predicting foraging behavior. This underscores the fundamental importance of energy to life, and its central role in the natural selection process that shaped the brains we have today.

OFT models the fundamental principles of food motivation that are wired into the brains of all humans. As it turns out, it has some interesting implications for those of us who hunt and gather, whether in the wild or in a supermarket.

PRIMAL INDULGENCES

The idea of moderation in eating is totally foreign to hunter-gatherers. In fact, Wood, Hill, and Pontzer explain that hunter-gatherer eating habits can be downright gluttonous. Hill recalls some of the enormous meals he

---

73 Hill: "If all that mattered to humans was energy, everybody would be farming corn or wheat, and that would be the entire economic activity of the world. In fact, what farmers actually do is they take their plant produce, and they feed it to animals in order to raise them at incredibly inefficient rates in order to produce high protein resources. The only way you could ever explain why humans waste their energy either raising animals or hunting animals in a farming community is that there must be something in meat that's lacking in plant foods that's a critical nutrient, and it's not energy."

observed among the Aché—men eating five pounds of fatty meat each in a sitting, drinking one and a half liters of pure honey, or eating thirty wild oranges similar to the fruit we buy in the grocery store. And it's not just the Aché. Pontzer adds that the Hadza also drink honey "like a glass of milk."

In contrast to our modern eating habits, the Hadza make every effort to extract the maximum number of calories from their food. As soon as they kill an animal, they pinch it in several places to determine how much fat it carries. They know exactly which parts are the fattest, and if they kill a large animal, such as a kudu or a zebra, explains Wood, they cut off the fattest parts, boil it down, and drink the soup. They break open and boil every bone until it's white and brittle, extracting every last gram of fat from its marrow. "They fully embrace the idea of 'eat as much pure fat as you can possibly eat,'" explains Wood, adding, "there's no hint of moderation whatsoever in their drive and motives with eating food." Hill's experience with the Aché echoes this: "Quite simply, they eat everything they can get and they don't seem to have limits."

Yet despite the fact that they guzzle sugar and fat when available, neither the Hadza nor the Aché have obesity. In fact, the Hadza have an ideal body composition by modern Western standards: men average 11 percent body fat, women average 20 percent, and neither gender becomes fatter with age. The only Hadza with obesity Wood has ever encountered was a rich man who didn't maintain a traditional diet or lifestyle. While Aché hunter-gatherers can have somewhat higher adiposity, especially young women, they rarely have obesity.

How can they engage in such gluttonous behavior and yet remain lean? The energy balance equation we encountered in chapter 1 allows only one possible answer: Their long-term average energy intake must match their energy expenditure.[74] The reality of hunter-gatherer life is that true starvation is rare, yet they often don't get as many calories as they would like. Despite appearing healthy and well fed, both Aché and Hadza adults

---

74 It's tempting to speculate that they're lean because they have such high levels of physical activity. Yet Pontzer's detailed metabolic studies show (surprisingly) that the Hadza don't expend any more energy in a twenty-four-hour period than the average semisedentary Westerner, when relevant factors such as body composition are taken into account. This implies that they're lean because they eat fewer calories than we do on average, not because they burn more calories trekking around Hadzaland.

frequently report feeling hungry. This isn't the kind of mild hunger pang that reminds us to saunter over to the fridge at lunchtime, but rather the powerful hunger a person feels when he hasn't eaten much at all that day. "When they say they're hungry," explains Wood, "it's a lot more meaningful."

In other words, instances of gluttony are balanced by periods during which they're eating less than they would like. Meat is often available,[75] but fatty meat gorges are uncommon. Honey is often available, but not usually in sufficient quantity to exceed a person's daily calorie needs. Simply stated, there isn't enough food to indulge their outsized appetites.

Why do hunter-gatherers report feeling hungry even when they appear to have enough body fat and muscle, and why do they wish they had just a bit more food? The answer may lie in the dynamics of hunter-gatherer reproduction. If life is a game of turning energy into kids, then more energy (up to a point) means more kids.[76] Since reproductive success drives natural selection, we might expect natural selection to have designed a brain that wants more energy. This is exactly what Hill thinks has happened: "Their brains are designed to want more food because more food converts into higher fertility and higher survivorship, and those things lead to higher [reproductive success]."

This leads us to a key conclusion about life as a hunter-gatherer: *Gluttony is good for them*. Eating as much sugar, fat, protein, and starch as possible, whenever it's available, increases their ability to thrive and bear children in a wild environment. "When those opportunities do present themselves," argues Wood, "there's basically no downside to going for it. It's all upside." Unlike in the modern affluent world, in which overeating is a major cause of ill health, in a hunter-gather environment, overeating is

---

75 It's fashionable at the moment to declare that hunter-gatherers are really near-vegetarian "gatherer-hunters" and therefore that vegetarianism, or something close to it, is the natural diet of humans. Yet a large body of evidence contradicts this idea. The most comprehensive analysis to date, including 229 present-day and historical hunter-gatherer cultures, shows that no known hunter-gatherer culture has ever been vegetarian, and while there is great variability between cultures, for most, animal foods are a major part of the diet. All the anthropologists I interviewed who have worked directly with hunter-gatherers confirmed this.

76 Of course, this logic breaks down if you take it too far, because obesity is a leading cause of infertility.

healthy. The same would probably have been true of our own ancestors until relatively recently in human history.[77]

For most of our history, our instinctive drives to seek large amounts of fat, sugar, starch, and protein were well aligned with our interests. There was no need to count calories or to feel guilty about eating too much. Yet in today's world of extreme food abundance, these same drives often undermine our health and even our ability to reproduce. We try to use the sophisticated tools of our cognitive mind to restrain our impulses to overeat, yet the impulses often win. The brain that drives hunter-gatherers to gorge on calorie-dense foods—because it's good for them—is the same brain that drives us to overeat in the modern world.

STALKING THE WILD CHICKEN NUGGET

Applying OFT to the hunter-gatherer environment and our own affluent world yields a striking contrast. In a hunter-gatherer environment, the calorie content of food items varies, but most foods require a substantial amount of effort and time to obtain and prepare. Therefore, in most cases, their overall economic value according to the OFT equation is relatively low. In other words, wild foods are usually a mediocre deal because they "cost" a lot. When high-calorie foods are easy to obtain, and therefore they have a very high value, hunter-gatherers take advantage of the situation by eating stupendous amounts of food, as we saw in the previous examples. These foods are a good deal because they deliver a lot of calories and don't cost much—and hunter-gatherers rarely pass up a good deal.

In the affluent world, we stalk Froot Loops, buffalo wings, and chicken nuggets rather than fruit, buffalo, and wild fowl. Most of our foods are rich in calories, and the time, effort, and monetary costs involved in acquiring and preparing them have drastically declined. If we apply OFT to this situation, it becomes clear that we're surrounded by an enormous variety of extremely valuable foods—foods that are an outstanding deal because

---

77 This is strongly supported by archaeological findings, which show that signs of periodic undernutrition during growth (Harris lines, enamel hypoplasia) were nearly ubiquitous among hunter-gatherer and agricultural populations. Undernutrition is a major cause of childhood mortality, in large part because it suppresses immune function and leaves children susceptible to deadly diseases.

they deliver a large number of calories and cost very little. Despite the fact that we live in a radically different environment than a hunter-gatherer, our brains are still highly attuned to good deals (have you ever seen how fast free pizza disappears?). Yet while a hunter-gatherer only encounters great deals occasionally, in our world, we encounter them multiple times per day. This leads us to overeat by activating largely nonconscious brain circuits that are constantly looking for a deal.

In Eric Ravussin's vending machine studies that we encountered in chapter 1, recall that his team provided volunteers with a variety of high-calorie foods that were free of cost, required almost no effort to prepare, and were readily available at all times of day. To eat food, the volunteers only had to wander into the next room and enter a code. Ravussin and his team unwittingly created a situation in which the value of food, according to the OFT equation, was exceptionally high—the ultimate deal. As predicted by OFT, this scenario resulted in spectacular overeating and rapid weight gain (opportunistic voracity, as Ravussin called it).

Brian Wansink, director of the Food and Brand Lab at Cornell University, has conducted clever experiments that illustrate the outsized influence of effort cost on our eating behavior. In one study, he recruited administrative assistants and placed candy dishes containing Hershey's Kisses in one of three different locations in each of their offices: on the desk, in the top drawer of the desk, or in a filing cabinet six feet away. Eating a Kiss on the desk required only a small arm motion, while Kisses in the drawer required a larger arm motion, and those in the filing cabinet required getting up and walking across the room. Each additional effort barrier, although small, made the Kisses a less attractive deal.

Remarkably, these seemingly trivial differences in effort cost resulted in large differences in candy intake. Participants with candy bowls on their desks ate an average of nine Kisses per day. Those with candy bowls in their desk drawers ate six Kisses per day, and those who had to hike all the way across the room only ate four Kisses. As Wansink puts it, "It's not worth the effort for an Eskimo to locate and overeat mangos." Would we be leaner as a nation if we all had to walk three miles and climb a tree each time we wanted a hamburger and fries (or ice cream, or pizza)? Almost

> ## STEP AWAY FROM THE SNACKS
>
> The practical implications for avoiding overeating are clear: Don't make it too easy for yourself to eat food throughout the day. Even effort barriers as small as having to open a cabinet, twist off a lid, peel an orange, or shell nuts can make the difference between eating the right amount and overeating. Keeping easy, tempting foods in plain sight, such as an open bag of chips or bowl of candy, creates a situation that is simply too tempting for the parts of our brains that are constantly on the lookout for a good deal.

certainly. Yet our food today is more convenient than it has ever been in human history.

YOUR BRAIN ON POP-TARTS

As I described in chapter 4, our food system has changed dramatically over the last century. Yet those changes don't just apply to the rewarding properties of foods but also to the costs of food, such as time, effort, and money, that determine their economic value to our brains. For example, between 1929 and 2012, US food spending dropped from 23 percent of disposable income to 10 percent (see figure 22). This alone makes food today a much better deal than it was when our grandparents were our age. The old adage says, "Hunger is the best sauce," but I would argue that "cheap" is a pretty good sauce too.

Food has become more convenient as the time and effort costs of obtaining and preparing it have dwindled. Grocery stores became common in the 1920s, providing us with the convenience of a single location for all our food shopping needs, and they have continued to expand in size since then. As each decade has passed, we've outsourced more of our food preparation effort to professionals in the restaurant and processed-food industry. And over the last fifty years, we've increasingly gravitated toward the ultimate

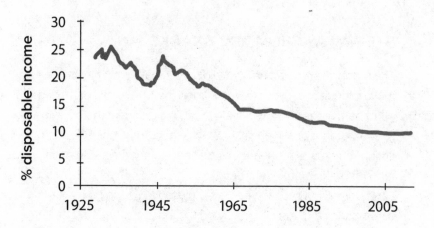

Figure 22. Percent of disposable income spent on food from 1929 to 2012. Data are from the USDA Economic Research Service.

convenience meal: fast food. Most of these restaurants focus on foods you can eat with your hands, dispensing with the dreadful inconvenience of silverware. We don't even have to leave our cars (or stop driving!) to eat anymore. (You can find a graph of home versus full-service restaurant versus fast-food restaurant spending over time in figure 17 of chapter 4.)

Even the workload in our own kitchens has declined. The food industry has responded to consumer demand for more convenient foods by creating zero-effort meals we can buy at the grocery store and eat at home. In his wonderful book *Salt Sugar Fat,* Michael Moss describes how the food industry giants carefully designed these products for maximum consumer appeal. Lunchables are premade lunch packs that allow busy parents to save the time and effort they might otherwise have to spend packing lunches for their kids. Pop-Tarts aim to replace breakfast with a thin, glazed pastry that heats up in the toaster. TV dinners and related meals only have to be heated in the microwave. The list of time- and effort-saving innovations developed by the food industry is endless. Innate human economic preferences created the demand, and the food industry was happy to meet this demand by inventing ultraconvenient food options.

Just as we engineered our food to satisfy our innate food preferences, humans have shaped our food environment to satisfy our innate economic preferences. Together, huge decreases in the time, effort, and monetary costs of food make many of our modern foods an exceptionally good deal. In the same manner that the economic preferences of the Hadza brain drive them to gorge when they encounter food that's a great deal, our own brains compel us to overeat in the same situation. The difference is that the Hadza only encounter great deals occasionally, whereas we encounter them multiple times per day. While many of these economic changes have improved our lives by freeing up time and money for other things, they have also contributed to our expanding waistlines by appealing to the innate economic logic of the human brain.

We're wired to seize a good deal when we see it, even at the expense of our waistlines. How does the brain recognize a good deal and motivate us to act on it?

THE VALUE CALCULATOR

In the laboratory of Camillo Padoa-Schioppa, associate professor of neuroscience and economics at Washington University in St. Louis, a rhesus monkey stares intently at a small black dot on a computer monitor. Suddenly, more shapes appear on the screen: a yellow square on the left and a blue square on the right. Then a black dot near each colored square. The monkey shifts its gaze to the black dot near the yellow box, and less than a second later, it receives a drop of grape juice through a tube in its mouth.

Padoa-Schioppa's research uses simple choice tasks in monkeys to understand how our brains compute cost-benefit decisions. In this particular trial, the monkey was offered a choice between one drop of grape juice (yellow square) and one drop of unsweetened Kool-Aid (blue square). The monkey indicated its choice by looking at the black dot near the yellow box, which, via repetition, it has learned represents grape juice (see figure 23A). Presumably because of its sweetness, rhesus monkeys almost always choose grape juice in this scenario.

In the next trial, the choices become a bit more complicated. The monkey is offered a choice between one drop of grape juice (one yellow box) and three drops of unsweetened Kool-Aid (three blue boxes) (see figure 23B). Now, there are two variables at play. Instead of only varying the *type* of juice, Padoa-Schioppa's team has also varied the *amount* of juice. In this

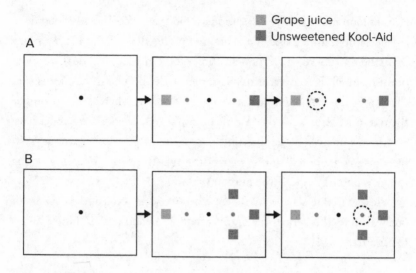

Figure 23. Camillo Padoa-Schioppa's economic choice tasks. In figure A, the monkey is offered a choice between one drop of grape juice (left) and one drop of unsweetened Kool-Aid (right). In figure B, the monkey is offered a choice between one drop of grape juice (left) and three drops of unsweetened Kool-Aid (right). The monkey makes its choice by directing its gaze to the dot closest to its selection. Adapted from Padoa-Schioppa et al., *Nature* 441 (2006): 223.

scenario, the monkey has to collect information about two variables for each option and decide which option it prefers overall. The monkey is a bit thirsty, so it looks at the three blue boxes and receives three drops of unsweetened Kool-Aid.

This second trial begins to hint at the complexity of everyday decisions. Some decisions are easy, such as choosing between one orange and two oranges. In this case, there is a clear quantitative difference. But what about a choice between an orange and an apple? What about a choice between a pastry behind a counter and three dollars in your wallet? We often choose between options that vary in many dimensions and have no objective means of comparison. Yet we're able to compare them anyway, and we often make seemingly rational decisions. How is that possible? How can we compare a pastry with three dollars if they have scarcely anything in common? There has to be a common unit that the brain uses to compare options on the same scale, and that unit is *subjective value*.

Subjective value quantifies how much each particular option will benefit the organism so different options can be compared on the same scale.[78] "Value is the only way to compare goods that are incommensurable, or qualitatively different," explains Padoa-Schioppa. Economists and psychologists have long known that people behave as if their decisions are guided by assigning subjective value to options, but only recently have neuroscientists begun to understand how the brain computes it.

While Padoa-Schioppa's monkeys make their choices, his team records the electrical activity of single neurons in a part of the brain called the orbitofrontal cortex (OFC). The OFC is a region of the prefrontal cortex, which is the part of the brain most often associated with reasoning and judgment (see figure 24). Relative to the rest of the brain, the prefrontal cortex is large in primates and even bigger in humans. Padoa-Schioppa has discovered that the firing patterns of individual OFC neurons represent the value of specific options.[79] For example, there are neurons that fire a little bit when a monkey is offered one drop of grape juice, and fire a lot when it's offered four drops. Then there are other neurons that fire in response to unsweetened Kool-Aid. And, interestingly, there are even neurons that represent the choice the monkey has made—before its eyes move to the target.

Padoa-Schioppa and other researchers have found that OFC neurons are able to integrate many kinds of cost-benefit information into their value computations, including the type of juice, its quantity, the probability of obtaining it, and the time and effort costs required to obtain it. Remarkably, the firing of these neurons seems to encode the complete subjective value of each option—in other words, each option's overall value to the monkey. Studies in humans have also implicated the OFC in value computation, as well as a nearby brain region called the ventromedial prefrontal cortex. The activity of these neurons may in fact be responsible for how much we value a pastry and how much we value three dollars.

To return to concepts we discussed in chapter 2, the OFC is reciprocally connected to the basal ganglia, suggesting that it may be an option

---

78 Or at least, how much the brain *thinks* an option will benefit the organism. Clearly, this value calculator sometimes leads to counterproductive choices in the modern environment.

79 I don't want to give you the impression that there is a single "grape juice neuron" in the OFC. There are many neurons that simultaneously respond to each option.

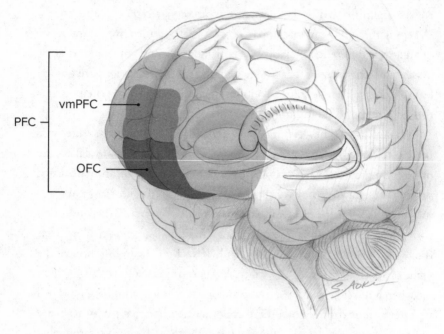

Figure 24. The prefrontal cortex, and within it, the orbitofrontal cortex and the ventromedial prefrontal cortex.

generator.[80] As a reminder, the basal ganglia sort through competing bids from option generators and select the most valuable one. This suggests the following scenario, which has not been directly demonstrated but is consistent with the evidence we have (illustrated in figure 25).

First, the OFC uses incoming information from other brain regions to calculate the predicted value of each option.[81] You independently compute the value of the pastry and the three dollars in your wallet. Second, the OFC sends those two independent bids to the basal ganglia, where the striatum compares them and selects the strongest one. You're feeling very tempted by the seductive food cues behind the counter, and the three dollars isn't worth much to you because you just got paid, so the pastry wins. Third, the basal ganglia return that selection to the

---

80 As is typical for option generators in the cortex, these connections arrive in the striatum and are relayed from the basal ganglia back to the prefrontal cortex via the thalamus.

81 For example, information from the hypothalamus and brain stem about how hungry you are, and information from the sensory cortex and brain stem about the appearance, smell, taste, and texture of the food.

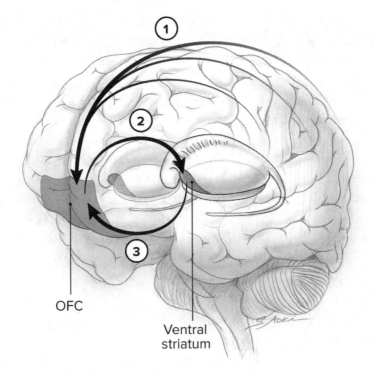

OFC

Ventral
striatum

Figure 25. The process of economic choice in the brain. First, the OFC uses information from many brain regions to calculate the value of each option. Second, it sends a bid for each option to the ventral striatum. And third, the striatum selects the strongest bid and returns it to the OFC (via other brain regions).

OFC, and the decision is made. This activates a cascade of other competitions in cognitive and motor areas of the brain, which skillfully plan and generate the behaviors necessary to follow up on the decision (see figures 11 and 12 on page 35). You reach into your wallet, pull out three dollars, trade them for the pastry, and dig in.

FAILURE TO COMPUTE

If the OFC really plays an important role in computing value, then disrupting its function should impair decision-making in specific ways. A person with OFC damage should still be able to act out well-worn habits, because they don't require value computations. If you always flush the toilet after using it, your brain doesn't have to compute the value of flushing versus not flushing. The decision had already been made when you stepped into the bathroom.

Yet OFC damage should reduce a person's ability to make decisions in response to changing conditions, because this requires the brain to compute value on the fly. And that's exactly what happens. This time, let's imagine you've just flushed the toilet. For most people, this new information would immediately reduce the value of flushing a second time, and so you wouldn't do it. Yet someone who has trouble computing value isn't able to incorporate that new information into her decision, so chances are, she'll flush again out of habit. And again, and again. This phenomenon, where a person keeps performing an action even after it's no longer useful, is called *perseveration,* and it often results from OFC damage.[82]

One common and striking consequence of OFC damage is overeating and weight gain. At first glance, this might seem strange. Why would damage to a brain region that computes an abstract concept like value affect eating so much? The answer may lie in the brain's inability to update the value of food as a meal progresses. When you sit down to a meal, that first bite of food usually has the highest value, because you're hungry. As the meal progresses, you feel increasingly full, and the value of each additional bite goes down. Once the value of the next bite is lower than the value of doing something else, like clearing the table, you stop eating. This process requires your brain to continually update the changing value of the food on your plate—precisely what people with OFC damage are unable to do. For a person with OFC damage, every bite is as compelling as the first one, so they often overeat substantially. This may explain why they continue to eat even when they report feeling extremely full: the brain registers the feeling of fullness just fine, but that feeling doesn't get translated into behavior because it doesn't register in the OFC.

---

82 Often due to frontotemporal dementia or stroke. You may recall that "the man who had no thoughts" due to basal ganglia damage in chapter 2 also showed perseveration. The reason is that since the OFC and the basal ganglia are reciprocally connected as a "loop" structure, damage to any part of the loop can cause similar deficits.

THE LAZIEST MICE IN THE WORLD

Now we know how the brain calculates value, but once it spots a good deal, how does it motivate you to pursue it? In chapter 2, we encountered the dopamine-deficient mice created by Richard Palmiter at the University of Washington. These mice are unable to produce dopamine naturally, and as a consequence, they sit nearly motionless in their cages, unable to eat or drink until their dopamine is chemically replaced. I explained that this is because dopamine lowers the threshold for selecting behaviors, and without it, the threshold is so high that nothing gets selected.

The research of John Salamone, professor of psychology at the University of Connecticut, allows us to refine this idea: Palmiter's mice may just be extremely lazy. So lazy, in fact, that even taking a few steps across the cage to eat and drink aren't worth the effort. How do we know? Because Salamone's research shows that dopamine plays a key role in motivation. When he reduces dopamine signaling in the ventral striatum of rodents, they become less willing to work for a reward. They choose easy options over hard options, even if the hard option has a much larger payoff and would normally be the better choice. In other words, reducing their dopamine signaling makes them lazy. Palmiter's mice have no dopamine, making them quite possibly the laziest mice in the world.

Dopamine seems to play a similar role in humans. Increasing dopamine levels using amphetamine makes people more willing to work for rewards, even if the reward is small or uncertain. Dopamine turns us into go-getters.

This makes perfect sense when we think about what's happening when the brain releases dopamine. Through repetition, your brain has learned to associate the sight, smell, sound, and other food cues with the reward of eating food. As soon as you see that bar of chocolate in the candy aisle, your dopamine level begins to spike. This burst of dopamine energizes you to grasp the chocolate and put it into your cart. If you had seen frozen green beans instead, you would have had a much smaller dopamine burst, proportional to the smaller reward, and you probably would have walked right past them. Dopamine tunes your motivation level to be proportional to the value of the reward you're pursuing—and it does so beyond your conscious awareness.

THE VALUE OF YOUR FUTURE SELF

If the brain was always rational in the way it computes value, few people would carry credit card debt, and no one would overeat. But we often make self-destructive choices, and this is particularly true when our decisions involve the future. As opposed to decisions that pit material objects against one another, like an apple versus an orange, we often make decisions that pit our present selves against our future selves. And the evidence shows that we often shortchange our future selves, with disastrous consequences.

Let's return to the example of a pastry versus three dollars. Most of us like the taste of pastries, yet we also recognize they're unhealthy. The benefit of eating a pastry is its reward value: We want it; we like it. This is a benefit you experience immediately as you bite into the pastry.

The *costs* of eating a pastry, on the other hand, are all incurred by your future self. For most of us, eating that pastry takes us one small step closer to obesity and ill health in the future. Also, what could you have done with those same three dollars next week? Would spending them take you slightly closer to defaulting on your rent or mortgage?

This is an example of how your nonconscious, intuitive brain competes with your conscious, rational brain. Your intuitive brain has no concept of the future and no understanding of abstractions like health and finances. It wants to eat things that taste good, *right now*. Your rational brain, on the other hand, understands the value of the future, and other abstract notions like obesity and money. It wants to protect you from the excesses of the intuitive brain; it wants your future self to be lean, healthy, and rich.

Which one wins? To a large extent, this depends on a psychological trait called *delay discounting*. This principle is illustrated by the famous 1970 "Stanford marshmallow experiment," in which children were offered a choice between getting one marshmallow now or two marshmallows in fifteen minutes.[83] The children were left alone with the marshmallow on the table during the fifteen-minute wait, so the temptation was intense.

---

83 Various iterations of the procedure used different rewards, such as pretzels, cookies, or pennies. The essential constant feature was that children were asked to choose between an immediate less preferred reward and a delayed more preferred reward.

Some children talked to themselves or covered their eyes in an effort to resist. And many popped the marshmallow into their mouths as soon as the researchers left the room. The children were essentially being asked to decide between a small, immediate reward and a larger, future reward. Which one they chose depended in part on how much they implicitly *discounted* (undervalued) the value of a future reward versus an immediate reward.[84] In other words, how much they valued their present selves versus their future selves. In the end, most of the children ate the marshmallow, forgoing the larger reward and shortchanging their future selves.[85]

Interestingly, a follow-up study showed that children who were better at delaying ended up slimmer thirty years later. In fact, for each additional minute they were able to delay eating the marshmallow as a child, they were slimmer by 0.2 BMI points as an adult. This means that a ten-minute difference in delay was associated with a fifteen-pound difference in adult weight. This makes sense: People who value their future selves highly will value long-term goals like leanness and health. For someone like this, two marshmallows fifteen minutes from now is worth nearly twice as much as one marshmallow right now, and a slimmer figure next summer may be worth more than a pastry right now. On the other hand, people who don't value their future selves very much may find that one marshmallow right now is worth more than two marshmallows in fifteen minutes, and that a pastry right now is worth more than a slimmer figure next summer. Multiple studies have confirmed that people who steeply discount the value of future rewards are more likely to have obesity. They're also more likely to be addicted to illegal drugs, alcohol, or cigarettes; more likely to gamble; and more likely to carry credit card debt.[86] In essence, these are all examples

---

84 The marshmallow test isn't a pure test of delay discounting, because it also measures the ability to exert willpower to suppress the desire to act on temptations. The technical term for this is *impulse control* (or *response inhibition*), and it's a component of the umbrella concept of impulsivity.

85 Some have criticized the interpretation of the Stanford marshmallow study, saying that the children may simply not have trusted the researchers to follow through on their promise of two marshmallows. While this could have played a role in the results, the findings remain broadly consistent with other more specific studies on delay discounting, suggesting that the result was at least in part due to individual differences in delay discounting.

86 Most of these studies are cross-sectional (i.e., they examined delay discounting in people who already have problems such as addiction). You might be tempted to wonder whether steeply discounting the future causes addiction, or whether addiction causes people to steeply discount the future. Both may be true, but a study by Janet Audrain-McGovern and colleagues (including Leonard

of pursuing immediate rewards without much regard for future consequences.

Delay discounting may seem irrational at first. Why would people be willing to seriously harm their future selves for an immediate reward as trivial as a pastry or another round at the slot machine? However, from an evolutionary perspective, it makes perfect sense—for one simple reason: The future is uncertain. Our species evolved in a dangerous environment in which we had about a 50 percent chance of living to age thirty-five. If you aren't certain you'll be alive next year, it's rational to value what's happening right now more than what might happen next year. In the environment of our ancestors, it was advantageous to evolve brains that intuitively value our present selves more than our future selves.

Yet in affluent countries today, the future is much more certain than ever before in human history. Mortality is far lower, and life expectancy far higher, than in the distant past. We have so much legal accountability in affluent countries that it's rational to hand over large sums of money to investment companies that grow it at a snail's pace, even if we won't be able to touch the money until we're retired! Today, it makes sense to value our future selves almost as much as our present selves—but the nonconscious brain regions that compute value and determine our motivations haven't caught up yet. This makes it all too easy to make choices that compromise our future finances, health, and weight, even if our intentions are good. And it goes a long way toward explaining why we overeat.

PICTURE THIS

Is there anything we can do to fight our natural tendency to shortchange our future selves? Research from Leonard Epstein's group and others suggests that the answer is yes—by giving the rational

Continued on next page

Epstein) showed that nonsmokers who steeply discount the future are more likely to take up cigarette smoking at a later date, suggesting that steep delay discounting does indeed contribute to addiction risk.

Continued from previous page

brain a little boost using an exercise called *episodic future think-ing.* This phrase is a complex-sounding term for a technique that's actually quite simple: Before making a decision, you imagine yourself in the future. When making a decision about something that pits your present self versus your future self, such as whether or not to eat a pastry, first imagine positive events in the future, such as your birthday or an upcoming vacation. Place yourself in the scene and imagine yourself enjoying it. The more vivid the imagery, the better. This process fires up the regions in your pre-frontal cortex that process abstract concepts like the future and therefore causes your brain to intuitively weight the future more heavily in its decision-making process. This attenuates delay dis-counting. Epstein's research shows that episodic future thinking reduces the intake of tempting, calorie-dense foods by nearly one-third in overweight women, and the same trick works for overweight children as well.

6

THE SATIETY FACTOR

At age fifty-seven, Elisa Moser was admitted to a hospital in Würzburg, Germany, with a collection of troubling symptoms. Her family members explained to the doctors that over the last three years, she had increasingly suffered from headaches, memory loss, poor vision, and childish behavior. Most strangely, over this same period Moser had developed "uncommonly extreme obesity."

Moser's condition continued to deteriorate, and four weeks after admission to the hospital, she died. Presumably due to the unusual nature of her case, a professor named Bernard Mohr decided to perform an autopsy on her body. In his autopsy notes, he noted that Moser's "abdomen of extreme dimensions" contained "uncommonly large fat deposits."

Mohr then examined her brain. As he lifted it out of her skull and turned it over to inspect its bottom surface, he noticed a tumor that had damaged her pituitary gland as well as the brain region immediately above it, the *hypothalamus* (see figure 26). The year was 1839.

Although Mohr couldn't have known it at the time, his finding may have been the first in a long line of research that helps explain why we overeat, why some people weigh more than others, and why weight loss is challenging and often temporary.

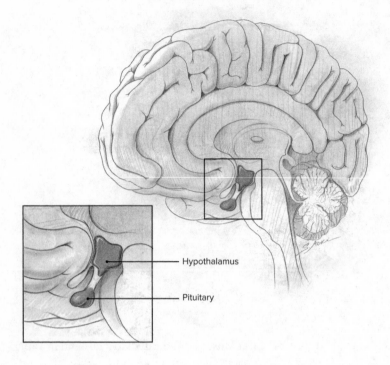

Figure 26. The hypothalamus and pituitary gland.

THE SEARCH FOR THE SATIETY CENTER

By the turn of the twentieth century, other researchers had begun to replicate Mohr's findings. In 1902, the Austrian American pharmacologist Alfred Fröhlich defined a cluster of symptoms, including obesity and sex hormone dysfunction, which were associated with brain tumors in the same location Mohr had described. This condition came to be known as *Fröhlich's syndrome.*

Initially, researchers attributed the obesity of Fröhlich's syndrome to a disruption of the pituitary gland, which at the time was already known to play an important role in growth and development.[87] This view predomi-

---

87 Via its secretion of growth hormone and the gonadotropins. We now know that the secretion of pituitary hormones is directed by the hypothalamus.

nated for three decades after Fröhlich's study was published, yet cracks in the hypothesis had already appeared shortly after Fröhlich's discovery. In 1904, the Austrian pathologist Jakob Erdheim reported that some patients with obesity had tumors located in the hypothalamus, above the pituitary, but with no apparent damage to the pituitary itself. When several research groups confirmed that damaging the hypothalamus but not the pituitary gland produces obesity in dogs and rats, the case was closed. The obesity of Fröhlich's syndrome was due to damage of the hypothalamus, not the pituitary.

Yet the scientific odyssey into the regulation of adiposity by the brain had only just begun. In the 1940s, Albert Hetherington and Stephen Ranson performed a series of studies that ushered in the modern era of obesity neuroscience research. They accomplished this by using a remarkable device called the stereotaxic apparatus, invented by British neurosurgeons at the beginning of the twentieth century. A stereotaxic apparatus—pictured in figure 27—fixes the skull in place and allows researchers (and neurosurgeons) to perform brain surgery in an incredibly precise and reproducible manner, which is why versions of it continue to be used today. Hetherington and Ranson adapted it for use in rats and quickly discovered that the critical location for obesity was not the hypothalamus as a whole but rather a subregion of the hypothalamus called the *ventromedial hypothalamic nucleus* (VMN).

As shown in figure 29, rats with VMN lesions became incredibly obese, with some exceeding two pounds.[88] Physiologist John Brobeck, who was the first to perform careful studies on the feeding behavior of these rats during his medical studies at Yale University in the early 1940s, described them as "ravenously hungry." They were so eager to eat that they began bingeing even before the anesthesia of their surgery had worn off. This initial binge would typically go on for several hours without interruption, and they would continue eating two to three times their normal food intake for the next month. Brobeck found that the degree of overeating was well correlated with the degree of weight gain and that restricting the rats to a normal food intake largely prevented weight gain. This led to the

---

88 As with many brain regions, there are actually two VMNs, one on the right side and one on the left. Pronounced obesity requires lesioning both sides.

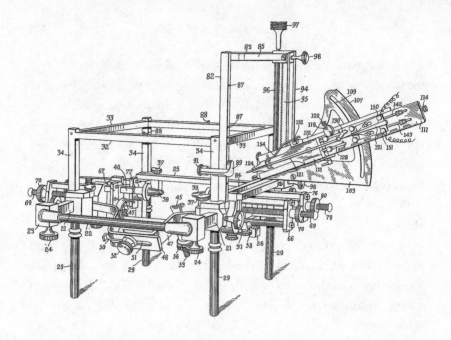

Figure 27. An early stereotaxic apparatus, patented in the United States in 1914 by Robert Clarke. The cranium is secured in the center of the apparatus by ear bars and a "mouth bit," and the probe on the right is used to perform surgery. Patent # US1093112-8.

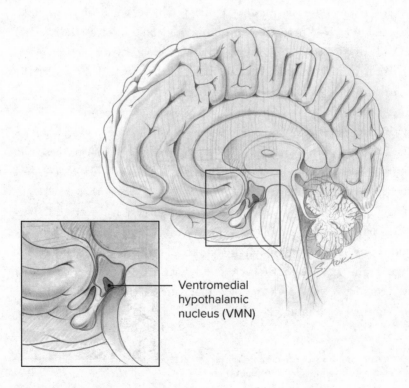

Ventromedial hypothalamic nucleus (VMN)

Figure 28. The ventromedial nucleus of the hypothalamus.

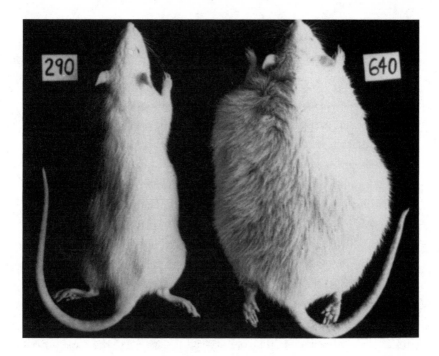

Figure 29. VMN-lesioned rat (right) next to normal rat (left). Numbers indicate weight in grams. Reproduced from P. Teitelbaum, *Proceedings of the American Philosophical Society* 108 (1964): 464, with permission of the American Philosophical Society.

conclusion that VMN-lesioned rats were obese primarily as a result of their excessive food intake.[89]

Researchers dubbed the VMN the *satiety center,* because disrupting it seemed to cause animals to lose the ability to feel full—and rapidly eat themselves to obesity. It was damage to this satiety center that caused obesity in Elisa Moser, in Fröhlich's patients, and in Hetherington's and Brobeck's rats.[90] Yet although they had identified its location, researchers still didn't know how the satiety center worked.

---

89 There was some early evidence that the obesity syndrome requires the sharp increase in insulin levels that follows VMN lesions. However, careful follow-up studies by Bruce King and colleagues later showed that increased insulin is not required for VMN-lesioned rats to become obese.

90 Today, it's widely accepted that tumors in the hypothalamus often cause excessive food intake and obesity in humans—hence the condition's modern name, *hypothalamic obesity.*

THE SEARCH FOR THE SATIETY FACTOR

In 1949—only a few years after Hetherington published his seminal VMN lesion studies—researchers at the Jackson Laboratory in Bar Harbor, Maine, puzzled over a mouse that appeared to be pregnant. The puzzling part was that the mouse never delivered a litter—and upon closer inspection, also happened to be a male. As it turned out, they had stumbled upon an obese strain of mice that had arisen due to a spontaneous genetic mutation. Dubbed the *obese* mouse, it was extremely fat and had an appetite to match (see figure 30). It also had a low energy expenditure for its size and metabolic disturbances reminiscent of human obesity. The inheritance pattern of its obesity suggested that the effect was caused by a single gene,[91] which they called the *ob* gene.

This was only the beginning of the era of obesity genetics research. In 1961, Lois and Theodore Zucker identified a strain of obese rats very similar to the *obese* mouse. The obesity-causing gene had the same inheritance pattern as the *ob* gene, and the rats became enormously fat, mostly (but not entirely) due to their prodigious appetites. In fact, they closely resembled rats with VMN lesions, with some animals exceeding two pounds. This strain was named the *Zucker fatty* rat. Over the ensuing decades, a number of other genetically obese rodent models would be identified, including *diabetes* and *agouti* mice. At the time these mutations were discovered, no one had any idea what the functions of the mutated genes were, or whether they were related to one another.

In 1959, just ten years after the *obese* mouse was identified, University of Leeds physiologist Romaine Hervey initiated a series of studies that would begin to uncover the cause of obesity in VMN-lesioned rats, *obese* mice, and *Zucker fatty* rats. To do this, he used a gruesome surgical technique called *parabiosis*. In parabiosis, two animals are surgically fused together at their flanks, effectively creating conjoined twins.[92] The key property of parabiosis is that it causes the circulatory systems of the two

---

91 *Recessive* inheritance pattern, meaning that both copies of a gene must carry the mutation for the animal to exhibit the trait.

92 The animals must be closely inbred so their immune systems don't reject one another.

Figure 30. The *obese* mouse. Right, normal mouse.

animals to exchange blood at a slow rate so that certain hormones released by one animal will affect the other. This allows researchers to determine if a phenomenon of interest—in this case, obesity—involves a circulating hormone.

To find out if the obesity produced by VMN lesions involved a hormone, Hervey lesioned the VMN of one parabiosed rat while leaving its twin intact. The results were striking: As expected, the VMN-lesioned rats became voracious and rapidly gained fat—yet unexpectedly, the non-lesioned twins lost interest in food, became emaciated, and often died of starvation.[93] Upon autopsy, the bodies of VMN-lesioned animals were overflowing with fat, while remarkably, their intact twins contained no visible fat at all.

For Hervey, the result suggested that a circulating factor was passing

---

93 You may be tempted to ask: If ingested calories are transported from the digestive tract to the tissues via the bloodstream, how is it possible for an animal to starve if its circulatory system is connected to an animal that's eating huge amounts of food and growing fat? The answer lies in the particulars of parabiosis. The exchange of blood is modest and slow, such that only powerful substances with a long half-life can exert effects in the other twin. As far as calories are concerned, the two animals are effectively independent.

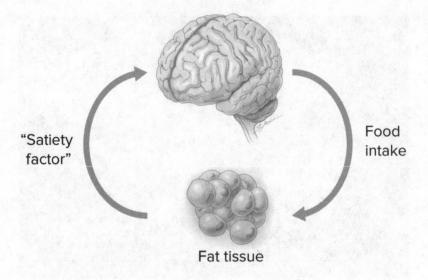

"Satiety factor"

Food intake

Fat tissue

Figure 31. Romaine Hervey's lipostat model. Fat tissue (bottom) secretes a satiety factor that acts in the brain (top) to constrain appetite and adiposity.

from the VMN-lesioned rat to its intact twin, suppressing the twin's appetite and adiposity. Building on a hypothesis recently developed by Gordon Kennedy, Hervey proposed that fat tissue secretes a hormonal *satiety factor* whose blood levels reflect adiposity, such that the more fat tissue a person carries, the higher his level of the hormone. This hormone then travels through the bloodstream and acts in the satiety center of the brain to constrain appetite and adiposity (see figure 31). The idea is that when body fat levels increase, the satiety factor increases, and this suppresses appetite and reduces adiposity back to its initial level. Conversely, when body fat levels decrease, the satiety factor decreases, and this stimulates appetite and fat gain back up to its initial level. Together, Hervey hypothesized, the satiety factor and the VMN form a feedback system that regulates adiposity, working to keep it at a stable level. They named this fat-regulating system the *lipostat,* after the Greek words for "fat" and "stationary."

Lesioning the satiety center of one parabiosed rat had caused its brain to become unresponsive to the satiety factor, making it think the rat was starving, and driving it to eat voraciously and become obese. In turn, Hervey reasoned, the lesioned rat's massive adiposity greatly increased the concentration of its now-ineffective satiety factor. Yet the intact twin was still responsive to the satiety factor, and as the hormone came pouring into its circulation from the obese twin, its satiety center received the signal and caused it to stop eating, become emaciated, and gradually starve.

In the early 1970s, at the Jackson Laboratory, a researcher named Doug Coleman set out to learn more about the mysterious *obese* mouse that had been identified there in 1949. He hypothesized that, like VMN-lesioned rats, *obese* mice have a defect in the lipostat system. Using parabiosis, Coleman joined *obese* mice with normal mice. In contrast to what Hervey had reported in his experiments with VMN-lesioned rats, the normal mice continued eating, and their weight remained stable. However, the *obese* twins underwent a remarkable transformation: Their appetite declined, they didn't grow as fat, and their obesity-related metabolic disturbances improved.[94] This led Coleman to conclude that *obese* mice lacked the elusive satiety factor and, therefore, that the satiety factor is encoded by the *ob* gene that is damaged in these mice. Joining them with normal mice had restored their damaged hormone, normalizing their appetite, body weight, and metabolism. Coleman published his findings in 1973, yet the identity of the *ob* gene remained shrouded in mystery.

Coleman's findings were a critical stepping-stone in the history of obesity research, because they gave us a defined entry point to unravel how the brain regulates appetite and body fatness. The *ob* mouse carries a mutation that inactivates the satiety factor hormone. If researchers could locate the gene, identify the hormone it produced, and discover how it works, they might be able to unlock the secrets of adiposity that had thus far eluded them.

Yet while Hervey and Coleman had shown that extreme disruptions of the lipostat, such as brain lesions and genetic mutations, can influence food intake and adiposity, no one had convincingly demonstrated that the satiety factor plays an important role in the normal, everyday regulation of food intake and adiposity. Ruth B. Harris, one of Hervey's former graduate students, set out to fill this knowledge gap in follow-up studies at the University of Georgia in the early 1980s. Harris and her team parabiosed two normal rats and overfed one of them using a feeding tube—similar to

---

94 Blood glucose and insulin levels declined toward normal values.

how farmers overfeed geese to make foie gras. The thinking was that if the satiety factor operates in normal animals, then as the overfed animal gains fat, it should produce more factor, and that should cause the twin to eat less and lose fat. As expected, as Harris's team overfed one twin and caused it to gain fat, the other twin lost fat. Further experiments showed that a small decline in food intake was required for fat loss to occur in the non-overfed twin. Again, this suggested that fat tissue (or something associated with it) was secreting a powerful hormone that constrains food intake and adiposity. Not only was this hormone relevant to animals with lesions and mutations but it was probably important for the everyday regulation of appetite and body fatness in normal animals.

Gradually, researchers converged on a remarkable conclusion: Even though each obesity model had been developed independently, VMN lesions, the *obese* mutation, the *Zucker fatty* mutation, and overfeeding all appeared to impact the same fat-regulating system. *Obese* mice are unable to produce the satiety factor; VMN-lesioned animals and *Zucker fatty* rats are unable to respond to it; and overfed animals overproduce it. These independent models all supported the fundamental importance of the same fat-regulating system Hervey had hypothesized in 1959: the lipostat.

Despite this profound convergence, Harris and other researchers still didn't know what hormone was responsible, as they had already ruled out every suspect that was known at the time.[95] This satiety factor was clearly central to the regulation of appetite and adiposity—yet no one had any idea what it was.

I 'm a baby doctor," explains Rudy Leibel in a slightly gravelly New York accent, "an endocrinologist baby doctor who got interested in how babies and children get obese, and decided to do something about it. I like that story. It happens to be true."

Leibel is currently an obesity researcher and professor of pediatrics at Columbia University. At the time of his medical training in the late 1960s

---

95 From a 2012 review paper by Harris: "In an effort to identify, or exclude, known hormones as potential 'satiety' factors we found no reliable changes in thyroid hormones, insulin, corticosterone, growth hormone, free fatty acids or ketone bodies of lean partners of overfed rats compared with members of control pairs." Other studies excluded the satiety hormone cholecystokinin and the fat-derived molecule glycerol.

and early 1970s, obesity research was still in its infancy, and this knowledge void allowed all sorts of harebrained theories to flourish. Obesity was often attributed to a slow metabolic rate or mysterious hormonal imbalances. Worse, it was frequently viewed through a psychoanalytical lens as a "neurosis, the physical expression of which is the accumulation of fat." At best, it was a moral failure resulting from gluttony or insufficient willpower.

Leibel was part of a growing contingent of researchers who were dissatisfied with these views. During his training at Massachusetts General Hospital, he became acquainted with *obese* mice, which didn't appear to require neuroses or moral failure to overeat and grow fat. In addition, he was familiar with a large body of evidence, gradually accumulated over the last century, suggesting that human body weight may be regulated—though it wasn't entirely clear by what. One of the earliest pieces of evidence came from Rudolf Neumann, a physiology researcher in Hamburg, Germany, who from 1895 to 1897 obsessively measured his own calorie intake and body weight. Neumann found that over a three-year period, his body weight remained surprisingly stable without any conscious effort to control it, despite the fact that his calorie intake fluctuated up and down in the short term. "That was probably the first exposure I had to the idea that there might be some very sophisticated regulatory mechanism for the maintenance or control of body weight in humans," recalls Leibel.

He had also been impressed by a number of studies suggesting that the human body vigorously resists large, short-term changes in weight brought about by underfeeding or overfeeding. One of the earliest and most influential studies was the Minnesota Starvation Experiment conducted by the prolific nutrition researcher Ancel Keys in the latter years of World War II. The goal was to understand the effects of starvation on the human body and mind. Over the course of six months of semistarvation, thirty-six young male conscientious objectors lost approximately one-quarter of their initial body weight. While this weight loss was no surprise, what happened after their food restriction was lifted is more interesting: Due primarily to a prodigious appetite, their body weights and adiposity rebounded rapidly. As they regained weight, their appetites normalized, and they eventually settled close to their original weights. It seemed as if a powerful internal control system was regulating their appetite and adiposity.

The system seemed to work in the opposite direction as well, pushing back against short-term fat gain. Obesity researcher Ethan Sims tremendously overfed a group of lean prison inmates in the 1960s, causing their weight to increase by up to 25 percent over a four- to six-month period. Despite the fact that these men were not particularly overweight even after overfeeding, their bodies vigorously resisted the weight gain. Sims found that he had to feed his subjects up to 10,000 Calories per day to get them to sustain their weight gain—nearly four times what most adult men require.[96] After the experiment was over, most of them hardly had any appetite for weeks afterward, and the majority slimmed back down to their former weights. This suggested, again, that an internal control system was regulating their appetite and body weight.

Leibel also knew about the research of Fröhlich, Hetherington, Ranson, Hervey, and Coleman, which suggested that the brain plays an important role in appetite and adiposity. And so in 1978, his curiosity got the better of him, and he took a position at Rockefeller University in New York City to look for the elusive satiety factor, under the tutelage of established obesity researcher Jules Hirsch. That decision cost Leibel an assistant professor appointment at Harvard University, and half his salary.

After Leibel spent several years investigating candidate factors without much success,[97] technical advances in genetics research promised to make it possible to identify disease-causing mutations in unknown genes. Today, this is a routine feat, but with the technology of the time, it was herculean, and success was far from certain. With Coleman's findings on the *obese* mouse in mind, Leibel explains, "I got to thinking that maybe we should try to identify one of these genes from a mouse." Researchers had already determined, due to its inheritance pattern, that the satiety factor hormone that was missing in the *obese* mouse was encoded by a single gene, the *ob* gene. The discrete genetic defect of the *obese* mouse offered a molecular entry point from which researchers could begin systematically unraveling the mysterious biology of appetite and body weight regulation.

Leibel had the concept and the drive, but he didn't have the technical

---

96 Some have questioned this 10,000 Calorie figure, and Sims himself takes a cautious view of it in later works. However, whether or not the number is exactly correct, the fact remains that they were eating substantially more calories than they should have required to sustain their weight.

97 Fatty acids, glycerol, and the ratio of the two.

skill. What he needed was a technical wizard who had mastered the rapidly evolving techniques of molecular biology.[98] This man was Jeff Friedman, a bright and driven assistant professor at Rockefeller. In 1986, Leibel and Friedman initiated a collaboration to "clone" the *ob* gene—in other words, to identify its location in the genome and its DNA sequence.

What followed was a grueling eight-year investigation, requiring more than four thousand mice and one hundred person-years of work. The result would shake obesity science to its foundation and eventually rebuild it into a more mature and sophisticated discipline.

As Leibel and Friedman's team narrowed in on the location of the *ob* gene, Friedman grew increasingly concerned that the more senior Leibel would receive the lion's share of the credit for the gene's discovery. Friedman insisted that Leibel stay out of the lab where the work was being conducted, although Leibel continued to play a guiding role in the project.

On December 1, 1994, without Leibel's knowledge, Friedman published the identity of the *ob* gene in the journal *Nature*. He reported that the gene codes for a small protein hormone that's secreted by fat tissue and circulates in the blood. Friedman named this hormone *leptin*, after the Greek word *leptos*, meaning "thin." Furthermore, he showed that humans carry an almost identical gene. The paper concludes, presciently:

> Identification of *ob* now offers an entry point into the pathways that regulate adiposity and body weight and should provide a fuller understanding of the [development] of obesity.

The satiety factor had been identified. And obesity was on its way to becoming a biological problem.

Leibel, and most of the other researchers who played key roles in the project, were excluded as authors on the paper. The day before the paper was published, Friedman had filed a patent for leptin. After a ferocious bidding war between pharmaceutical companies, that patent was sold to

---

98 In particular, a collection of techniques called *positional cloning,* which permits the mapping and sequencing of unknown genes in unknown locations.

Amgen for a $20 million initial payment so it could develop leptin as the ultimate weight-loss drug.[99]

AN INCREDIBLE DRIVE TO EAT

The identification of leptin triggered a feeding frenzy among researchers and the pharmaceutical industry. The race was on to understand how leptin works and to explore its potential as a weight-loss drug.

The first step was to purify leptin and see how administering it to rodents affects their adiposity. Echoing Coleman's parabiosis findings, Friedman and his collaborators showed that leptin injections curbed the prodigious appetite of *obese* mice and caused them to slim down, exactly as predicted. Leptin precisely plugged the physiological hole that drove them to gain fat. Most intriguingly—especially for the pharmaceutical industry and the public—injecting large doses of leptin into normal mice caused their body fat to melt away almost completely, without affecting their muscle mass. Just as Kennedy, Hervey, and Coleman had predicted over the last half century, leptin was a hormone produced by fat tissue, which acts in the brain to regulate appetite and adiposity.

That is, in rodents. The significance of leptin in humans remained unclear until a remarkably lucky finding by Stephen O'Rahilly, professor of clinical biochemistry and medicine at the University of Cambridge, in 1996. O'Rahilly studies the genetics of diabetes and obesity, sifting through countless cases of these disorders to look for individuals who truly stand out—"clinical exceptions" who just might carry genetic mutations that shed light on the mechanisms of disease. Shortly after Friedman's team published their paper, O'Rahilly was searching for humans with leptin mutations, and he had a pair of candidates: two cousins of Indian descent, both of whose parents were first cousins in the same family (inbreeding increases the likelihood that a rare mutation will occupy both copies of a gene, resulting in a genetic disorder). One of the cousins weighed 189 pounds

---

99 Because of Friedman's authorship of the *Nature* paper, and his unilateral patenting of leptin, he is thought to have received the large majority of the profits that were distributed to the research team by the university (the patent holder). This controversy is covered in greater detail in Ellen Ruppel Shell's excellent book *The Hungry Gene*.

by the age of eight and had to use a wheelchair to get around. Liposuction had failed to attenuate her extreme horizontal growth. The other cousin weighed 64 pounds by the age of two. Both children had an extraordinary obsession with food from a very young age, and seemingly insatiable appetites. To O'Rahilly, this wasn't garden-variety obesity, and it had nothing to do with psychology. Something was seriously wrong with the biology of these children.

It was Sadaf Farooqi's first month as a clinical fellow in O'Rahilly's lab, and her first assignment was to look for leptin in the two cousins. She was unable to detect it. Suspecting that she might have run the experiment incorrectly, she repeated it. Again, the children showed undetectable levels of leptin. O'Rahilly and Farooqi had found human counterparts of the *obese* mouse on their first try.

Eventually, Farooqi and O'Rahilly were able to show that the children's insatiable appetites and extraordinary obesity were caused by the omission of a single guanine nucleotide—one letter in the 3.2 billion-letter genetic code—that happened to inactivate the leptin gene. "That was really the first evidence that a defect in a single human gene could cause obesity," explains Farooqi, "but also that a complete lack of leptin could cause obesity in humans."

At this point, Farooqi and O'Rahilly have worked extensively with leptin-deficient humans, who are extremely rare. "Usually, they are of normal birth weight," says Farooqi, "and then they're very, very hungry from the first weeks and months of life." By age one, they have obesity. By age two, they weigh fifty-five to sixty-five pounds, and their obesity only accelerates from there. While a normal child may be about 25 percent fat, and a typical child with obesity may be 40 percent fat, leptin-deficient children are up to *60 percent* fat. Farooqi explains that the primary reason leptin-deficient children develop obesity is that they have "an incredible drive to eat," resulting in an exceptionally high intake of calories.[100] In addition, the reward regions of their brains show an exaggerated response to images of calorie-dense, high-reward foods. Leptin-deficient children are nearly always hungry, and they almost always want to eat, even shortly after meals. Their appetite is so exaggerated that it's almost impossible to put them on a

---

100 They may also have a low energy expenditure, although this has been hard to prove.

diet: If their food is restricted, they find some way to eat, including retrieving stale morsels from the trash can and gnawing on fish sticks directly from the freezer. This is the desperation of starvation.

Furthermore, leptin-deficient children have a powerful emotional and cognitive connection to food. "They really enjoy food," explains Farooqi, "and if you provide them with food, they're incredibly happy. It doesn't matter what the food is." Even the dreaded hospital cafeteria doesn't faze them. Conversely, they become distressed if they're out of sight of food, even briefly. If they don't get food, they become combative, crying and demanding something to eat.

Unlike normal teenagers, those with leptin deficiency don't have much interest in films, dating, or other teenage pursuits. They want to talk about food, about recipes. "Everything they do, think about, talk about, has to do with food," says Farooqi. This shows that the lipostat does much more than simply regulate appetite—it's so deeply rooted in the brain that it has the ability to hijack a broad swath of brain functions, including emotions and cognition, and put them to work seeking food.

There's another condition that causes similar behavior: starvation. Let's return to the Minnesota Starvation Experiment for a moment, but this time, let's focus on the psychological responses that occurred. Over the course of their weight loss, Keys's subjects developed a remarkable obsession with food. In addition to their inescapable, gnawing hunger, their conversations, thoughts, fantasies, and dreams revolved around food and eating—part of a phenomenon Keys called "semistarvation neurosis." They became fascinated by recipes and cookbooks, and some even began collecting cooking utensils. Like leptin-deficient adolescents, their mental lives gradually began to revolve around food. Also like leptin-deficient adolescents, they had very low leptin levels due to their semistarved state.[101]

The striking similarity between leptin deficiency and starvation suggested that low leptin levels might be responsible for the human brain's response to starvation, including the hunger, the obsession with food, the activation of brain reward regions, and, as we will soon see, the reduced

101 To be clear, Keys didn't measure leptin because it was unknown at the time. But we know from later work that semistarvation and weight loss cause large reductions in leptin levels.

metabolic rate. It seemed as if the absence of leptin in leptin-deficient children prevented their brains from "seeing" their fat, triggering a powerful starvation response despite the fact that they had extreme obesity.

This phenomenon of starvation in the face of abundant body fat is a paradox with which Rudy Leibel and Jules Hirsch are very well acquainted. In 1984, they published a seminal paper showing that people with garden-variety obesity who then lose weight show evidence of a starvation response. Beginning with a group of twenty-six people with an average body weight of 336 pounds, Leibel and Hirsch slimmed them down to 220 pounds using a rigorous low-calorie diet. Although 116 pounds is an impressive amount of weight to lose, at the end of the weight-loss period, the volunteers were still considered to have obesity. Remarkably, after weight loss, the number of calories their bodies consumed was only about three-quarters of what it should have been based on their new, slimmer body size. What's more, they were ravenously hungry. Something was shutting down their metabolic rate and ramping up their appetites—fighting their weight loss—even though they remained quite fat.

Leibel and Hirsch have spent much of their careers following up on this puzzling finding. In further studies, they found that weight loss, both in people who are lean and obese, triggers a powerful suite of biological and psychological responses that work together to restore the lost fat. To do this, the brain curtails the activity of the sympathetic nervous system and reduces thyroid hormone levels, both of which slow the metabolic rate, accounting for the cold and sluggish feeling some people experience after weight loss. The brain cuts back the number of calories a muscle burns during a given contraction, reducing the calories expended in physical activity. And most important, the brain ramps up hunger and increases the response to food cues that signal high-calorie, high-reward foods. Before weight loss, you might have been able to stroll by the ice cream aisle without any problem, but after weight loss, the temptation to buy and eat ice cream can be overwhelming. In effect, substantial weight loss triggers a starvation response, whether a person is lean, overweight, or obese—and this response continues until the fat comes back.

If you've never had the experience of fighting your own body's starvation response, Jeff Friedman provides a helpful analogy:

Those who doubt the power of basic drives, however, might note that although one can hold one's breath, this conscious act is soon overcome by the compulsion to breathe. The feeling of hunger is intense and, if not as potent as the drive to breathe, is probably no less powerful than the drive to drink when one is thirsty. This is the feeling the obese must resist after they have lost a significant amount of weight.

Friedman's analogy is an important lesson for people who think weight loss is as easy as deciding to eat less and exercise more. The brain was forged in the flames of highly competitive natural selection, and it doesn't take weight loss lightly. "[The starvation response] is there for preservation," says Leibel. "Evolutionarily, it has played a very important role in our survival."

Remarkably, Leibel and Hirsch found that this starvation response could be almost completely eliminated by injecting their volunteers with small doses of leptin, just enough to keep leptin levels where they were before weight loss. This shows that the drop in leptin that occurs with weight loss is the key signal that triggers the starvation response. Although this powerful self-preservation mechanism evolved to keep us alive and fertile, it often seems to backfire in the modern affluent world, where excess adiposity is a greater threat than starvation.

The findings of Leibel, Hirsch, Friedman, O'Rahilly, Farooqi, and many others lead us inescapably to the conclusion that appetite and adiposity are, to a large extent, biological phenomena that are regulated by nonconscious parts of the brain. Their research has brushed away a variety of archaic hypotheses that were blissfully unconstrained by the need for supporting evidence, and left us with a clear understanding that food intake and adiposity are not simply the result of conscious, voluntary decisions. Yet the leptin story also delivered a major disappointment.

DETECTING DEFICIENCY

Because of a tiny genetic error that disrupted the leptin gene, the brains of O'Rahilly and Farooqi's patients were unable to perceive their vast fat stores, and as a result, triggered the ultimate starvation response. No amount of food was able to satisfy these children whose brains thought they were in the final, deadliest stage of starvation.

Fortunately for the leptin-deficient cousins, O'Rahilly and Farooqi obtained approval to treat them with injections of isolated leptin. The effect was immediate and dramatic. While before leptin treatment no amount of food was enough, within four days of treatment, they began turning down food. Their obsession with food subsided too, and their brain responses to tempting foods normalized. They lost much of their excess fat, and after a few years, they looked and acted similar to normal kids.

This raises a $20 million question: Why aren't we all taking leptin to lose weight?

It turns out that people with garden-variety obesity—as opposed to obesity caused by a rare genetic mutation—already have high levels of leptin. And researchers have found that leptin isn't the miraculous obesity cure the pharmaceutical industry hoped it would be. While leptin therapy does cause some amount of fat loss, it requires enormous doses to be effective (up to forty times the normal circulating amount). Also troubling is the extremely variable response, with some people losing over thirty pounds and others losing little or no weight. This is a far cry from the powerful fat-busting effect of leptin in rodents. The new miracle weight-loss drug never made it to market.[102]

This disappointment forced the academic and pharmaceutical communities to confront a distressing possibility: The leptin system defends vigorously against weight loss, but not so vigorously against weight gain. "I have always thought, and continue to believe," explains Leibel, "that the leptin hormone is really a mechanism for detecting deficiency, not excess." It's not designed to constrain body fatness, perhaps because being too fat is rarely a problem in the wild. Many researchers now believe that while low leptin levels in humans engage a powerful starvation response that promotes fat gain, high leptin levels don't engage an equally powerful response that promotes fat loss.

Yet *something* seems to oppose rapid fat gain, as Ethan Sims's overfeeding

---

102 As a weight-loss drug, anyway. It is currently FDA approved to treat the metabolic problems that accompany lipodystrophy, a rare disorder in which body fat atrophies and leptin levels plummet, causing metabolic problems.

studies (and others) have shown. Although leptin clearly defends the lower limit of adiposity, the upper limit may be defended by an additional, unidentified factor—in some people more than others. We'll return to this in the next chapter. For now, let's explore how this fat-regulating system works and what the implications are for those of us who would like to lose weight or stay lean.

THE FAT THERMOSTAT

The leptin system functions by the same principle as the thermostat in your home, which measures the ambient temperature and compares it to the temperate set point you've programmed. If the temperature dips too low, your thermostat engages the heating system; if it rises too high, it engages the air conditioner. This feedback system serves to maintain the stability, or *homeostasis,* of your home's interior temperature.[103]

The body maintains homeostasis for a number of variables, including temperature, blood pressure, blood pH, respiration, and pulse rate. These variables are regulated because they're so important for survival.

And so, like your home thermostat, the brain maintains temperature homeostasis by measuring skin and core body temperature and taking action to heat or cool the body as necessary (see figure 32). This includes a variety of physiological and behavioral strategies like constricting blood vessels in your skin to reduce heat loss or dilating them to increase it, revving up a special heat-producing tissue called brown fat, making you shiver to generate heat, or prompting you to put on a sweater or gravitate toward shade and ice water. This coordinated strategy is so effective that it manages to keep your core temperature within about one degree Fahrenheit, regardless of the weather.

As it turns out, the body's thermostat resides in the hypothalamus. It receives temperature information from sensors in the body and coordinates the physiological and behavioral responses necessary to maintain ideal body temperature. Similarly, the hypothalamus (and other brain regions,

103 This is an example of a "negative feedback" system, in which deviations from a set point are met with an opposing reaction that brings the regulated variable back toward the set point.

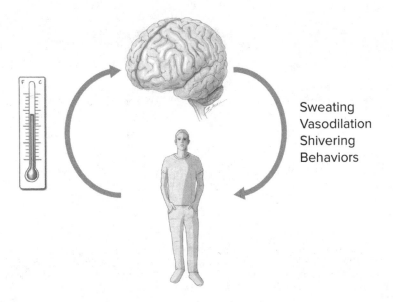

Figure 32. Temperature homeostasis. The brain (top) measures body temperature using internal and external thermometers (left), and regulates body temperature accordingly by using a variety of physiological and behavioral strategies (right).

to a lesser extent) is the body's lipostat—the brain region that regulates appetite and body fatness. It receives information about the size of fat stores from signals, including leptin, and coordinates the physiological and behavioral responses necessary to maintain adiposity (see figure 33). As Leibel and Hirsch observed in their weight-loss studies, if a person loses fat, the lipostat engages a coordinated suite of responses that work to increase energy intake, reduce energy expenditure, and thereby regain the lost fat. This is a milder form of the same starvation response Farooqi and O'Rahilly described in leptin-deficient children. Yet the analogy with a thermostat isn't perfect: In humans, the lipostat isn't as good at preventing fat gain, as if your home thermostat has very good heat to prevent the temperature from dropping, but weak air conditioning to prevent the temperature from rising. As you may have noticed, our modern understanding of the lipostat closely resembles the model that Romaine Hervey proposed in 1959 (figure 31).

Leibel and Hirsch's findings suggest that the lipostat isn't broken in garden-variety obesity—it simply regulates adiposity around a higher *set*

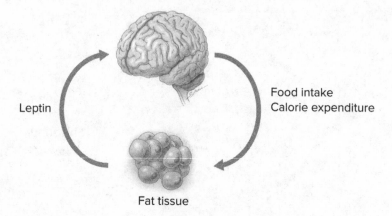

Leptin

Food intake
Calorie expenditure

Fat tissue

Figure 33. The lipostat. The brain (top) measures adiposity using leptin and other signals (left), and regulates adiposity accordingly by using a variety of physiological and behavioral strategies that influence food intake and energy expenditure (right). In humans, food intake is the primary means by which the brain regulates adiposity.

point, analogous to turning up the thermostat in your home.[104] When the adiposity set point is high, the brain requires a higher level of leptin to restrain the starvation response, and in the long run, the only way to get more leptin is to have more fat. In other words, for the brain of a person with obesity, obese is the new lean. Researchers call this phenomenon *leptin resistance,* because the brain seems to have a hard time "hearing" normal levels of leptin.

What are the implications for us? The first is that once a person develops obesity, it becomes a self-sustaining state, and the person *has to overeat* to feel the same satisfaction that a lean person feels after eating a smaller meal. In essence, once we gain weight, the lipostat becomes one of the primary reasons why we continue to overeat, undermining our conscious desire to be lean and healthy.

A second key implication is that weight loss is hard because it requires us to fight deeply wired impulses. Long-term diet trials suggest that the hypothalamus is remarkably good at undermining fat-loss efforts. Weight

104 There has been vigorous academic debate over whether or not the brain regulates adiposity using a true unitary set point, rather than a "settling point" that's the net output of several interacting systems. For our purposes, the distinction isn't important, but I wanted to take a moment to acknowledge that this debate is ongoing.

regain afflicts all the most popular diet styles, including portion control, low-fat diets, and low-carbohydrate diets. One of the best examples of this comes from the hit TV series *The Biggest Loser*. On this reality show, contestants with obesity adopt an extreme diet and exercise regimen, and the person who loses the highest percentage of her starting weight at the end of the season wins a cash prize of $250,000. Many contestants lose over 100 pounds of body weight, including Ali Vincent. In 2008, Vincent dropped 112 pounds, securing a win in the show's fifth season. Yet since her nadir of 122 pounds, she has regained most of the weight. "I feel like a failure," says Vincent, understandably frustrated by gradually losing the war of attrition against her hypothalamus. Her experience is by no means unique. Suzanne Mendonca, who lost 90 pounds in the show's second season in 2005, quipped, "NBC never does a reunion. Why? Because we're all fat again."

The hypothalamus doesn't care what you look like in a bathing suit next summer, and it doesn't care about your risk of developing diabetes in ten years. Its job is to keep your energy balance sheet in the black, and it takes that task very seriously because it was essential for survival and reproduction in the time of our distant ancestors. The tools at its disposal, including hunger, increased food reward, and slowing your metabolic rate, are extremely persuasive. When the hypothalamus squares off with the conscious, rational mind, it usually wins in the end. This doesn't mean dieting is hopeless, but to be successful, it's important to understand, respect, and work with what you're up against. The good news is that the lipostat responds to the cues we give it through our diet and lifestyle, and we can use this to our advantage.

APPEASING THE LIPOSTAT

If the brain regulates adiposity, then how does a person go from being lean to overweight or obese? And can this process be reversed? We know that in each instance of homeostasis, such as heart rate, body temperature, and adiposity, the body defends a certain set point value against changes—but that doesn't mean set points are inflexible. For example, your body temperature set point can increase in response to infections, a phenomenon we call fever. When you have a fever, your brain doesn't lose control of temperature regulation—it deliberately regulates it around a higher set point to fight the infection. Your body turns up its thermostat. Similarly, the set point of the lipostat can be turned up and, some evidence suggests, turned down.

Leibel's and Hirsch's weight-loss studies lend scientific rigor to a common-sense conclusion: Different people defend different adiposity set points. In lean people, the lipostat defends a low-adiposity set point against fat loss, which makes great sense from an evolutionary standpoint—a lean person can't afford to lose much fat. Yet what makes less sense is that among people with obesity, the lipostat defends a high-adiposity set point. Somehow, the hypothalamus has "decided" to defend an obese body type rather than a lean one, even though a person with obesity carries much more fat than is necessary to avoid starvation and infertility. In fact, it makes even less sense than that, because excess adiposity is one of the leading *causes* of infertility and premature death in the affluent world.

Many researchers, including myself, speculate that the lipostat behaves abnormally because it's been placed in an unfamiliar situation: The hypothalamus is wired to keep us healthy and fertile in an environment that disappeared long ago. In today's environment of plentiful, tasty, refined, calorie-dense food, low physical activity requirements, and other unnatural characteristics, the lipostat essentially misfires, driving many of us inexorably to overeat and gain weight. Yet others seem to remain lean no matter what they do—a topic we'll return to in the next chapter.

The set point doesn't just differ by individual; it can change over time in the same person. Most people in affluent nations gain fat over the course of a lifetime, showing that the set point can move upward, gradually increasing the lower limit of our comfortable weight. This malleability explains how a lean nation like the nineteenth-century United States can become dangerously overweight over the course of a few generations, without a significant change in genetic makeup. Our body weight isn't completely determined by our genes. Just like the temperature set point, the adiposity set point responds to the conditions of our lives.

In 2000, Barry Levin, an obesity and diabetes researcher at Rutgers University, published a paper clearly demonstrating this effect in rats. Starting with a genetically diverse strain of rats, he fed them either ordinary rat pellets or a high-calorie palatable diet. On the palatable diet, some of the rats gained weight and fat, while others didn't. Levin's team took the rats that had gained weight and restricted their food intake while keeping them on the palatable diet, which caused them to lose weight and fat. So far this is what you might expect, but what they found next is more interesting: When they lifted the calorie restriction, allowing the animals to eat as much as they wanted again, the rats didn't just resume their gradual upward trajectory of weight gain—they quickly bounced back up to the weight of rats that had been eating an unrestricted palatable diet and gaining weight the entire time. Levin's findings suggest that there is a certain weight that the lipostat "wants" an animal to maintain, defending it against changes, and that this weight depends both on genetics and on the specific diet the animal is eating.

Levin's team went further, testing the effects of rotating an obesity-susceptible strain of rats between three different diets. The first diet was ordinary rat pellets, the second diet was the same palatable diet as the previous experiment, and the third diet was a highly palatable milkshake-like meal-replacement beverage called Ensure[105] (specifically, the chocolate flavor). As expected, the rats began overeating and gaining weight and fat on the palatable diets. In fact, the rats eating Ensure almost doubled in weight over ten weeks—an impressive feat.

Yet when Levin's team made the rats switch diets, they found that the rats seemed to defend dramatically different body weights depending on which diet they were currently eating. For example, when they were switched from Ensure to ordinary pellets, their food intake dropped impressively and they rapidly lost weight, until they approximated the weight of animals that had been eating pellets the whole time. When these same animals were switched back to Ensure, they gorged and quickly bounced back up to the weight of animals that had been eating Ensure the whole time. Again, it seemed as if the diets were not just passively causing weight gain but actually changing the set point of the lipostat. Levin ascribed much of this effect

---

105 Yes, this is the same stuff they give to elderly people who have trouble eating enough and need a nutritional boost.

to the diet's palatability, in part because the rats would only overeat and gain weight on chocolate-flavored Ensure—not vanilla or strawberry![106]

   This effect hasn't been studied as thoroughly in humans, but there are tantalizing clues that it might apply to us as well. Let's return to the studies we encountered in chapter 3, in which volunteers with obesity lost weight rapidly on an unrestricted bland liquid diet dispensed by a machine through a straw. The volunteers were specifically instructed to drink as much of the liquid diet as they needed to feel satisfied, yet once they began the diet, their calorie intake spontaneously dropped because they simply weren't very hungry (even though lean people on the same regimen drank a normal amount of calories, showing that it wasn't physically difficult to do so). They lost weight rapidly, yet their starvation response never seemed to kick in. Something about the bland diet was allowing their bodies to feel comfortable at a lower weight, suggesting that just as with Levin's rats eating plain old rat pellets, the low reward value of the diet may have lowered their adiposity set point.

   Five years after the machine-feeding studies, Michel Cabanac, a physiology researcher at Laval University in Canada, published another study that supported and expanded upon these findings. Cabanac's team had a group of volunteers eat an unrestricted bland liquid diet for three weeks, which caused them to spontaneously eat fewer calories and lose just under seven pounds. Then the researchers had a second group of volunteers lose the same amount of weight over the same period of time by deliberately restricting the portion size of their normal diet. Cabanac found that the portion control group developed the expected hunger response to weight loss—but the bland diet group didn't. He reported that the bland diet volunteers "reduced their intake voluntarily and were always in good spirits," while the portion control group "had to continually fight off their hunger and would spend the night dreaming of food." On the bland diet, the starvation response never kicked in. Cabanac concluded that diet palatability influences the set point of the lipostat in humans.

   Returning to the brain, we know that there are important connections between the hypothalamus and reward regions such as the ventral stria-

---

106 In case you're wondering, all three varieties of Ensure were nutritionally identical—the only thing that differed was the flavoring.

RESTRICTING REWARD

High-reward foods tend to increase food intake and adiposity, while lower-reward foods tend to have the opposite effect. This suggests a weight management "secret" you'll rarely find in a diet book: eat simple food. The reason you'll rarely find it in a diet book is that, by definition, lower-reward food is not very motivating. It doesn't get us excited about a diet, and it doesn't make books fly off the shelves. We want to hear that we can lose weight while eating the most delicious food of our lives, and the weight-loss industry is happy to indulge us. The truth is that there are many ways to lose weight, but all else being equal, a diet that's lower in reward value will control appetite and reduce adiposity more effectively than one that's high in reward value. The trick, as with all diets, is sticking with it, because just as the set point can go down, it can go right back up if you return to your former eating habits. And that means designing an eating plan you can live with for the long haul. For most people, a "bland liquid diet" like those I described isn't a viable long-term solution, but keeping added fats, sugars, salt, and calorie-dense highly rewarding foods to a modest level may be.

tum, because hunger magnifies food reward. This is encapsulated in the old adage "Hunger is the best sauce." Yet we don't know as much about the mechanisms underlying the reverse connection: how food reward might affect the brain regions that determine appetite and adiposity.

We do, however, have enough information to arrive at some practical conclusions. First, calorie-dense, highly rewarding food may favor overeating and weight gain not just because we passively overeat it but also because it turns up the set point of the lipostat. This may be one reason why regularly eating junk food seems to be a fast track to obesity in both animals and humans. Second, focusing the diet on less rewarding foods may make it easier to lose weight and maintain weight loss because the lipostat doesn't fight it as vigorously. This may be part of the explanation for why all weight-loss diets seem to work to some extent—even those that are based on diametrically opposed principles, such as low-fat, low-carbohydrate, Paleo, and vegan diets. Because each diet excludes major reward factors, they may all lower the adiposity set point somewhat.

Are there other ways to appease the lipostat? Researchers have often noted that people who exercise more frequently gain less weight over time. There's a seemingly straightforward explanation for this: They're burning more calories, so they remain in energy balance. This is probably at least partially correct, but there may be more to the story, as Barry Levin's research suggests. His findings show, not surprisingly, that exercise attenuates weight gain in rats when they're offered a fattening diet.[107] Yet Levin's data also reveal that fit rats aren't just leaner—they actively defend a lower adiposity set point than sedentary rats on the same diet. This is actually quite consistent with human studies, in which physically fit people are better able to resist fat gain in the face of overeating. It appears that exercise helps keep the lipostat happy at a lower set point.

Yet many people have pointed out that exercise doesn't work very well as a weight-loss tool in humans, and they have plenty of data to back them up. If you send volunteers home with advice to exercise regularly, most of them hardly lose any weight. Superficially, this presents a striking contrast to the rodent studies. However, I've gradually come to believe that there's more to this story than meets the eye. The problem with many of the human studies is that they simply offer people exercise advice, without having any way to enforce the advice, and often without even accurately measuring how much exercise was actually performed.

In contrast, when we only consider studies in which volunteers had to regularly report to a research gym and exercise under supervision—ensuring compliance—a different picture emerges. In these studies, fat loss is often substantial, and it increases with the intensity and duration of the exercise regimen. So it appears that many of us in the research world, including myself at one time, may have misjudged exercise: It really does cause fat loss.

But the evidence is more nuanced than it may initially appear. The

---

107 A refined diet higher in fat and sugar than normal rat chow.

research of John Blundell, professor of psychobiology at the University of Leeds, shows that not everyone loses the same amount of fat as a result of exercise. Intrigued by earlier observations that people respond differently to exercise, Blundell and colleagues recruited thirty-five overweight and obese sedentary men and women and assigned them to exercise five times per week for twelve weeks. Each session was designed to burn 500 Calories and was supervised by the researchers to ensure compliance. At the end of the twelve-week period, the average participant had lost more than eight pounds of fat. However, this average conceals some very interesting information: Changes in adiposity ranged from a loss of twenty-one pounds to a *gain* of six pounds! To be fair, only one person out of thirty-five gained fat, and we don't know what else was happening in that person's life at the time, but it does show that fat gain is possible in the face of a vigorous exercise program. Two others lost less than a pound of fat—a paltry reward for so much effort.

How is it possible to expend 2,500 extra Calories a week yet gain fat? Again, our answer must lie in the energy balance equation we encountered in chapter 1. The only way for adiposity to increase despite increasing calorie expenditure is if calorie *intake* increases by an even greater amount. And that's exactly what Blundell's team observed. When they measured calorie intake in the volunteers, they found that those who lost less weight than expected were inadvertently increasing their calorie intake in response to exercise. This isn't particularly surprising, since most of us have probably had the experience of "working up an appetite" after playing sports or doing yard work. What's remarkable is what happened in people who lost as much, or more, weight than expected: They actually *decreased* their calorie intake in response to the exercise regimen. In the end, about half of the volunteers ate more as a result of the exercise, and half didn't.

Presumably, this reflects the effects of exercise on the lipostat, as suggested by Levin's studies in rats. On one hand, exercise depletes the body's fat reserves and therefore triggers the lipostat to increase appetite. On the other hand, exercise may lower the adiposity set point among people who carry excess fat, *reducing* appetite and facilitating fat loss. The strength of each of these opposing forces, which varies from person to person, determines the net change in appetite in response to exercise. So while exercise can cause substantial fat loss, it works better for some people than others.

Another fact that's often overlooked is the difference between weight loss and fat loss. When someone tries to slim down, the goal is not usually to lose weight—it's specifically to lose fat. It turns out that exercise helps preserve muscle mass during weight loss. Although slow progress on the scale may be frustrating, changes in the mirror and in health as a result of exercise can be better than what the scale suggests.[108]

In the end, the evidence suggests that if you can maintain a high level of physical activity, it will probably help you prevent fat gain, accelerate fat loss, and maintain that loss. But it only works if you actually do it—and even then, the degree of fat loss depends on how effectively your brain compensates for the lost calories by increasing your appetite.

Low-carbohydrate dieting is one of the most popular ways to lose weight, and numerous studies suggest that although it's no miracle obesity cure, it is more effective than a traditional low-fat portion-controlled diet over periods of about a year. This is actually a big deal, because it's a major reversal of the view that's prevailed for most of the last half century, which is that fat is fattening, and the best way to lose weight is to cut back on fat. In fact, many people report that low-carbohydrate diets help them curb their appetite and cravings, and the research backs this up. When people go on a low-carbohydrate diet, their spontaneous calorie intake drops substantially—even though they usually aren't making any deliberate effort to eat fewer calories.

Why? As you may have noticed, this effect looks a whole lot like what happens when the adiposity set point goes down. If we take a closer look at the diets of people who eat a low-carbohydrate diet, what we see is that when they reduce their carbohydrate intake, the proportion of protein in the diet tends to go up. As it turns out, amino acids, the building blocks of protein, act directly in the hypothalamus, influencing the lipostat system. Although most of the direct evidence comes from rodent studies, a substantial amount of indirect evidence suggests that a high intake of protein may be able to lower the adiposity set point in humans too.

---

108 Physical activity also improves health independently of changes in body composition. So even if you aren't gaining muscle or losing fat, exercise is still good for you.

A 2005 study by University of Washington researcher Scott Weigle and colleagues illustrates just how striking this effect can be. Weigle's team started by determining the habitual calorie intake of a group of nineteen volunteers, after which the researchers fed them a high-protein diet (30 percent of calories) for twelve weeks under tightly controlled conditions. On the high-protein diet, the volunteers' calorie intake spontaneously declined by an average of 441 Calories per day, and their weight dropped by nearly eleven pounds—despite the fact that it wasn't a weight-loss study and no one had asked them to eat less. As expected, their leptin levels declined as they lost weight—yet the starvation response never seemed to kick in. The effect can't be attributed to reducing carbohydrate intake, because Weigle's team increased protein at the expense of fat, not carbohydrate.

Adding to the case is the work of Maastricht University researchers Klaas Westerterp and Margriet Westerterp-Plantenga, which supports the idea that high-protein diets attenuate the starvation response that typically undermines weight loss. Their studies show that people who lose weight by eating a high-protein diet experience less hunger than those who lose weight by other means, and extra protein also largely prevents the reduction of energy expenditure that often accompanies dieting. Consistent with this, their findings demonstrate that restricting carbohydrate without increasing protein doesn't cause the same weight-loss effect as a typical higher-protein, low-carbohydrate diet, suggesting that carbohydrate restriction per se isn't actually the key ingredient in low-carbohydrate diets.[109] Rather, advice to eat a low-carbohydrate diet may be effective simply because it's an easy way to get people to eat high-protein foods and reduce major food reward culprits.

The lipostat system Bernard Mohr first stumbled upon in 1839 is a key nonconscious regulator of food intake and body fatness that often drives us to overeat. It helps explain why weight loss is so hard and why our appetites and waistlines react in certain ways to the cues we give them through our food and lifestyle. But we have much left to uncover as we

---

109 This may not apply to more stringent "ketogenic" low-carbohydrate diets, in which extreme carbohydrate restriction shifts the body into a unique metabolic state. There is some evidence that these diets may suppress appetite through mechanisms that don't depend on increasing the proportion of protein in the diet.

delve deeper into the machinery of the lipostat. In the next chapter, we'll explore how genetics affects the lipostat, why some people can eat as much as they want and not gain fat, and how a related system in the brain stem affects how many calories you eat at a sitting.

7

THE HUNGER NEURON

started my fellowship in 1987," reminisces obesity researcher Mike Schwartz about his training at the University of Washington, "and was immediately indoctrinated into the idea that there is an adiposity control system." Schwartz's fellowship was part of a very unusual research program led by obesity and diabetes researchers Dan Porte and Steve Woods. "What I was working on was considered way out of the mainstream at the time," explains Schwartz. "Everyone just assumed that obesity is a problem where people eat too much, and if they could control their eating and be normal, they wouldn't have this problem."

Although some scientists at the time believed that adiposity is regulated, most didn't. When Schwartz began his fellowship, Leibel and Friedman hadn't yet identified leptin, and researchers knew very little about the lipostat. Schwartz's goals were to raise awareness about the fact that adiposity is biologically regulated and eventually target it therapeutically. The only way to accomplish these two goals was to understand the underlying brain systems. Schwartz's research, and that of many others, would eventually explain much about how the brain regulates body fatness, how the lipostat changes in the brain of a person with obesity, and why some people are particularly prone to overeating and fat gain.

Three years before the start of Schwartz's fellowship, a researcher named Satya Kalra discovered that a small protein called neuropeptide Y (NPY) causes massive overeating when it's injected into the brain of a rat. Adding to the excitement, researchers discovered that NPY is naturally produced by neurons in the arcuate nucleus, a tiny region of the hypothalamus near the VMN satiety center, and it became more abundant after fasting, suggesting that it could be involved in hunger (see figure 34). Together, this led Schwartz, Porte, and Wood to hypothesize that NPY might be part of the brain apparatus that regulates eating and adiposity.

At the time, Schwartz's research was focused on the hormone insulin, which plays an important role in regulating the levels of sugar and fat in the blood.[110] Because of the fact that circulating insulin levels increase when a person overeats and gains fat, and decrease when a person under-eats and loses fat, Schwartz viewed it as a possible signal to the brain that participates in regulating adiposity. His team was able to show that injecting insulin into the brains of rats reduces the production of NPY in the hypothalamus and also reduces food intake. This was the first time anyone had been able to draw a biological road map from food intake, to a circulating hormone, to a brain circuit, and back to food intake.

Yet, explains Schwartz, "we knew insulin wasn't enough." They were well aware of Coleman's parabiosis work in *obese* mice, and also aware that insulin couldn't explain his findings. Something else had to be out there—something much bigger.

When Friedman's team published the identity of the *ob* gene, Schwartz, Porte, and Woods immediately realized leptin might be the missing link

---

110 Certain low-carbohydrate diet advocates have claimed that insulin's primary role is to regulate fat storage in fat tissue and that eating carbohydrate increases insulin levels and traps fat inside fat cells, causing fat gain. While this idea is appealing by virtue of its simplicity, it's difficult to reconcile with the modern scientific understanding of the biology of insulin and energy balance. Insulin does regulate the dynamic meal-to-meal flux of fat into and out of fat cells, but this function doesn't appear to impact the amount of fat that ends up being stored at the end of the day. Despite high insulin levels, people with obesity release fat from their fat cells at a higher rate than lean people, showing that neither insulin nor anything else is trapping fat inside their fat tissue. If insulin drives weight gain, then people with higher insulin levels should tend to gain weight faster than people with lower insulin levels, but this isn't typically the case. The reality is that adiposity isn't primarily regulated at the level of fat tissue—it's primarily regulated at the level of the brain, which in turn controls what's happening to fat tissue.

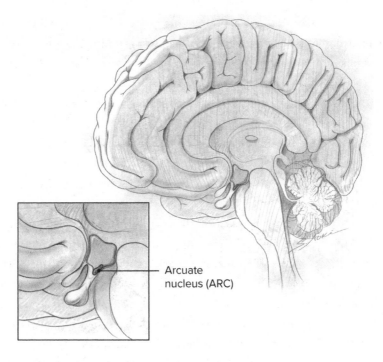

Arcuate
nucleus (ARC)

Figure 34. The arcuate nucleus of the hypothalamus.

they were looking for. "It seemed logical to speculate that leptin might do something similar to insulin," recalls Schwartz, "which was to inhibit neurons that stimulate feeding." After four months of grueling work, and thousands of microscope slides, the answer was no longer speculation: Leptin reduced hunger-promoting NPY levels exactly as predicted, supporting the idea that leptin controls food intake (in part) by reducing NPY levels in the brain. His finding was one of the first steps toward understanding how the lipostat works.

Schwartz submitted his team's findings to the journal *Science*—the highest-impact scientific journal in the world. "I was just beginning my assistant professorship," explains Schwartz. "No one knew who I was, so this was an important opportunity." A month after submission, he received a mailed letter from an editor at *Science,* the contents of which are permanently seared into his memory: "It has come to our attention that a paper has been accepted for publication elsewhere that significantly compromises the novelty of your findings." A competing group, led by Mark Heiman at the pharmaceutical company Eli Lilly, had scooped him. Schwartz ended up publishing his team's version of the finding in the journal *Diabetes,* which is a good journal, but not as visible as *Science.* This setback galvanized Schwartz and his team. "We had been burned, and there was a whole series of things we knew were going to be a race. We ended up winning most of those races because we knew what was at stake."

What followed was a virtual avalanche of data from Schwartz, Porte, Woods, Woods's new postdoc Randy Seeley, and several other competing groups. Shortly after the NPY study, Schwartz published a paper showing that the hypothalamus, and particularly the arcuate nucleus, contains high levels of the receptor for leptin. Even more tantalizing was the accumulating evidence that another group of proteins, called melanocortins, play the opposite role of NPY in the brain: When injected into the brains of rodents, melanocortins powerfully suppressed food intake.[111] Like NPY, melanocortins are located in distinct neurons of the arcuate nucleus, which were named POMC neurons, after the POMC protein that is the precursor of melanocortins. Schwartz's group showed that melanocortin levels are also regulated by leptin—yet in the opposite direction of NPY. NPY and melanocortins are key cellular pathways by which leptin regulates food intake and adiposity via the brain.

What emerged from these studies was a remarkably logical explanation for how leptin regulates the lipostat: It turns off neurons that drive eating, and it turns on neurons that inhibit eating. And, by implication, when leptin levels decline, neurons that drive eating turn on and neurons that inhibit eating turn off, increasing the drive to eat. This "push-pull" system is redundant and extremely robust, and only disrupting major nodes in the signaling pathway can derail it.

Disrupting a major node is precisely what Albert Hetherington and Stephen Ranson did by damaging the VMN "satiety center" in the rat studies we encountered early in the last chapter. The VMN contains neurons that stimulate POMC neurons, and when it's destroyed, POMC neurons become less active and cease to restrain appetite. As Hetherington and Ranson showed, this makes rats overeat tremendously and grow to impressive

---

111 Particularly a melanocortin called α-melanocyte-stimulating hormone (α-MSH).

proportions. But it doesn't require brain damage to disrupt a major node in the signaling pathway. Getting rid of leptin is another way to do so, and as we've seen, the result is the same.

Schwartz's basic explanation for how leptin works has withstood the test of time, and NPY and POMC neurons have remained at the center of the story, although we now know a lot more about how the system operates. For example, NPY neurons in the arcuate nucleus don't just secrete NPY; they secrete at least two other substances that stimulate feeding.[112] Because of this synergy of appetite-stimulating substances, released in just the right downstream brain regions, NPY neurons are the most powerful driver of eating known to science. If there is such thing as a "hunger neuron" that drives pure, visceral hunger, the NPY neuron is it.[113]

Scott Sternson, a neuroscience researcher at the National Institutes of Health's Janelia research campus, knows this well. His group was the first to specifically stimulate NPY neurons in awake, normally behaving mice.[114] When they turn on NPY neurons, mice eat. A lot. I've replicated this experiment myself, and it's quite impressive. With the flick of a switch, a mouse will stuff its face with whatever food is around—eating up to ten times what it would normally eat in the same period of time.

Furthermore, Sternson's work shows that the way NPY neurons compel a mouse to seek food is by making the mouse feel bad until it eats.[115] Like humans, mice don't like to feel hungry, and relieving that hunger state by eating—or, in Sternson's case, by turning off NPY neurons directly—is itself a reward. If we tie this together with what we covered in chapter 3, it becomes clear that eating motivates us in two distinct ways that reinforce one another: Unpleasant hunger neurons get turned off, and food reward neurons get turned on.

112 The first is agouti-related peptide (AgRP), which blocks the ability of melanocortins to activate POMC neurons. The second is gamma-aminobutyric acid (GABA), the principal inhibitory neurotransmitter of the brain.

113 These neurons are more commonly called *AgRP neurons* or *NPY/AgRP neurons* in the research world, because NPY also occurs in neurons outside the ARC that don't regulate adiposity, whereas AgRP doesn't. For simplicity's sake, I'm calling them NPY neurons here.

114 Using optogenetics, a technique we encountered in chapter 3.

115 I'm using the phrase "feel bad" loosely here. We don't know how a mouse really "feels," but we do know that hunger and NPY/AgRP neuron activation are negative reinforcers in mice (i.e., mice will learn to avoid both states).

Richard Palmiter, the University of Washington researcher who keeps popping up in this book, has a neat trick up his sleeve: He can destroy almost any neuron population in the brain with pinpoint accuracy, without harming nearby neurons.[116] When he uses this trick to destroy NPY neurons in *obese* mice, their appetites normalize, they lose weight, and ultimately, they become difficult to distinguish from normal mice. "The major symptoms are all corrected," explains Palmiter. What this suggests is that the primary reason *obese* animals overeat and become tremendously fat is that their NPY neurons are in constant overdrive because there isn't any leptin around to keep them in check. Get rid of the NPY neurons, and the animals slim down, even without a trace of leptin. This has a second, even more remarkable implication: The hunger, the obsession with food—many of the physiological and psychological effects that we see in people who are dieting, starving, or born without leptin—could be largely due to a population of hyperactive NPY neurons that is small enough to fit on the head of a pin.

At this point, there has been an enormous amount of research on the brain systems that regulate appetite and adiposity—far too much for me to summarize in this book. There are many other hormones and neurons that play roles in the system. Yet you don't have to know all those details to have a basic understanding of how the lipostat works. Think of it as an hourglass, with NPY and POMC neurons at its narrow center (see figure 35). On the top of the hourglass, we have signals entering the brain that communicate the body's current energy status. They include things like leptin and insulin. These signals converge, mostly through indirect routes, on NPY and POMC neurons, and collectively, they determine the activity of these neurons.

The bottom of the hourglass represents the outputs of NPY and POMC neurons. These are the responses that the brain uses to regulate the body's energy status, such as hunger, food reward, metabolic rate, and physical activity.

As far as we currently know, NPY and POMC neurons are the most

---

116 First, he uses genetic techniques to express the diphtheria toxin receptor, which mice naturally lack, specifically on NPY (more precisely, AgRP) neurons. Then, the mice are injected with diphtheria toxin, which kills the neurons but doesn't harm other cell types.

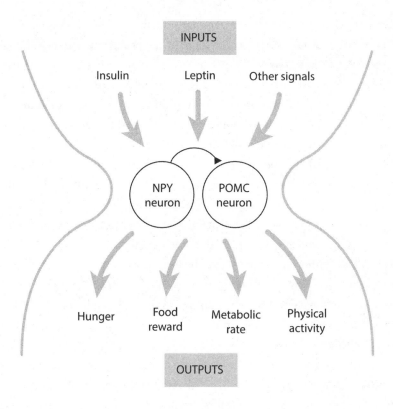

Figure 35. Regulation of adiposity by neurons in the arcuate nucleus. Top, inputs to NPY and POMC neurons. Bottom, outputs of NPY and POMC neurons. As depicted, NPY neurons contact POMC neurons. When activated, the former inhibit the latter, sweeping aside a major satiety mechanism.

important convergence points between the adiposity-regulating inputs and outputs of the brain, and as such, they have attracted a disproportionate amount of attention from the scientific community. Many researchers, including Sternson, Palmiter, and Brad Lowell, a neuroscience researcher at Harvard Medical School, are working on deciphering the inputs and outputs of these neurons—and they are making remarkable progress. "The ability to identify individual neurons and then use this technology to do circuit mapping," says Schwartz, "is ultimately going to take the field to the next level."

In many ways, we've already reached the next level Schwartz is referring to. In fact, we've cured obesity countless times—in rodents. We now have the ability to take genes from almost any species, manipulate them to make them do what we want, insert them into the mouse genome so that they are expressed in specific cell populations of the brain, and use them to influence food intake, adiposity, and many other things. We can precisely activate, silence, or even kill specific populations of neurons in the mouse brain, controlling appetite and adiposity as if the mouse were a marionette. Modern neuroscience has achieved feats that would have seemed like science fiction to researchers only a few decades ago.

We know from the work of Mohr, Leibel, Friedman, O'Rahilly, Farooqi, and many others that the brain circuits that regulate eating and adiposity in humans have much in common with those of rodents. With time, we could undoubtedly adapt the techniques I described for use in ourselves. So what's holding us back from curing human obesity? In a word, *ethics*. While we already have the technical ability to genetically engineer humans, and probably directly manipulate the brain circuits that control eating, we don't currently consider it ethical. And there are many good reasons for this, one of which is that we haven't yet demonstrated the safety of this technology.

We do consider it ethical to use drugs to nudge the brain circuits that control eating, and researchers have already developed many weight-loss drugs that do precisely that. Unfortunately, drugs are an extremely blunt tool, poorly suited for targeting an organ as complex as the brain. This is because when we take a pill or an injection, it bathes the entire brain in the drug, including all eighty-six billion neurons, and trillions of connections between them, which together perform countless unique tasks. Adding to the challenge, most of the chemical signals that regulate eating and adiposity are used for other purposes elsewhere in the brain and body. This makes it very difficult to target your circuit of interest without collateral damage. Imagine trying to drive nails into your wall with a sledgehammer: you may get a nail in sometimes, but you're also going to put holes in your drywall. Because of this, most drugs that impact food intake have unacceptable side effects, such as the dangerous psychological effects of the drug rimonabant ("reverse marijuana") we encountered earlier. Very few of the weight-loss drugs identified so far have acceptable side effects, and the few that do aren't the silver bullets we wish they were. But we keep looking, hoping for a lucky break, and it's possible that our expanding knowledge will someday make that possible.

At this point, Schwartz thinks the field has accomplished one of his goals: raising awareness about energy homeostasis. There is little remaining doubt among researchers, and even many doctors, that appetite and body fatness are biologically regulated by nonconscious regions of the brain. As far as his second goal, Schwartz hopes that our greater understanding of the lipostat will soon fulfill its promise to prevent and reverse obesity in humans. For me, this goal doesn't seem very far off, but, as Schwartz cautions, "that's what I thought when they discovered NPY."

Although this work explains a lot about how the lipostat works under normal conditions, it doesn't tell us what changes cause the brain of a person with obesity to defend a higher level of adiposity, or how we might reverse those changes. For that, the field would need a different approach.

SCARY IMPLICATIONS

Licio Velloso, an obesity researcher at the University of Campinas in Brazil, is determined to understand the changes in the brain that underlie obesity. In the early 2000s, he decided to take a new approach to tackling the problem, one that wasn't constrained by existing ideas about what might be happening in the brain. To do this, Velloso turned to a technique called an RNA microarray that tells the researcher which genes are turned on, which are turned off, and to what degree. By looking at the pattern of gene expression, we can peer into the inner workings of the cell and understand, to some extent, what it's up to at the moment.

Velloso wanted an answer to a simple question: What are the cells of the hypothalamus doing when an animal becomes obese? To do so, he used an RNA microarray to compare gene expression in the hypothalami of lean rats versus rats made obese by diet. When Velloso's team analyzed the data, a striking trend emerged: Many of the genes that were more active in obese mice were related to the immune system, and particularly a type of immune system activation called *inflammation*. As Velloso noted in his 2005 paper, this makes perfect sense. Previous research had already implicated chronic inflammation in insulin resistance—a condition in which tissues like liver and muscle have a harder time responding to the glucose-controlling hormone insulin—and this process had already been linked to increased diabetes risk. It wasn't a major leap to suppose that inflammation in the hypothalamus might cause resistance to leptin and insulin, increasing the adiposity set point and contributing to obesity risk.

To further test this idea, Velloso's team blocked a major inflammatory pathway in the brains of obese rats.[117] They reasoned that if inflammation in the hypothalamus is really causing obesity, then blocking this inflammation should reduce food intake and body weight. And that's exactly what they observed. Since Velloso's discovery, other researchers have followed up on the finding, confirming that inflammation in the hypothalamus blocks leptin signaling, leading to leptin resistance and weight gain.[118]

Yet inflammation isn't the only thing going wrong in the hypothalami of obese rodents. In 2012, my colleagues Josh Thaler and Mike Schwartz and I published a study in which we looked more closely at the cellular changes that occur in the hypothalamus during the development of obesity.[119] Much of the study focused on two types of cells in the brain called *astrocytes* and *microglia*. While neurons are the cells that do most of the information processing of the brain, astrocytes and microglia play a supporting role in keeping delicate neurons happy—protecting them from threats, helping them heal, giving them energy, and cleaning up after them.[120] When the brain is injured, these cells go into overdrive, increasing in size and number to counter the threat and accelerate healing. "All conditions that damage the brain," explains Thaler, "such as trauma, stroke, neurodegenerative disease, even infections to some extent, cause this effect."

In a healthy brain, astrocytes are small cells that send out a web of thin filaments to monitor surrounding cells, and these filaments don't overlap with the filaments of neighboring astrocytes. In an injured brain, astrocytes multiply and grow in size, and their filaments enlarge and overlap

---

117 C-Jun N-terminal kinases.

118 A large part of this effect may relate to the fact that inflammation activates a protein called SOCS3, which inhibits the activity of the leptin receptor. Mice that can't mount a normal inflammatory response in the brain, or which lack SOCS3, are genetically resistant to developing obesity when they're fed with a fattening diet.

119 I was a postdoctoral fellow in Schwartz's lab at the time, and Thaler was lead author of the study.

120 Some astrocyte researchers would bristle at this statement, because astrocytes do seem to participate in information processing in some situations. However, it's clear that neurons are by far the main information processing cells of the brain.

Astrocytes

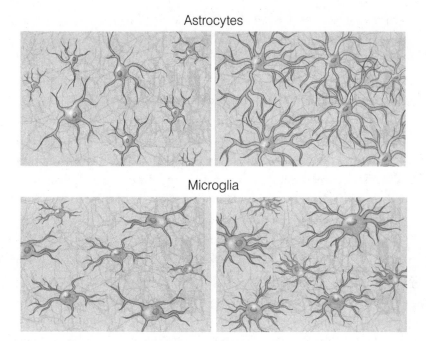

Microglia

Figure 36. Astrocytes and microglia in their resting (left) and activated (right) states.

those of neighboring astrocytes (see figure 36). Microglia undergo similar changes. The activation of microglia and astrocytes is a universal marker of brain injury, and it's visible under a microscope, so we decided to look for it in the hypothalami of our obese rats and mice.

And we found it: Astrocytes in the hypothalami of obese rats and mice were enlarged, and their filaments were tangled together in a thick mat. Microglia had also enlarged and multiplied. Both changes were specifically located in the same area as NPY and POMC neurons (the arcuate nucleus), but not elsewhere. Our results suggest that obese rodents suffer from a mild form of brain injury in an area of the brain that's critical for regulating food intake and adiposity. Not only that, but the injury response and inflammation that developed when animals were placed on a fattening diet *preceded* the development of obesity, suggesting that this brain injury could have played a role in the fattening process.

This is pretty interesting stuff if you're a rat, but what does it have to do with humans? To see if humans with obesity show evidence of injury in the hypothalamus, we called on our colleague Ellen Schur, an obesity researcher at the University of Washington. She specializes in a technique called *magnetic resonance imaging* (MRI), which allows researchers and doctors to observe the structure of live tissues without harming them, sort of like an X-ray that can see soft tissues in great detail.

One of the conditions doctors use MRI to diagnose is brain damage resulting from a past injury, such as stroke or physical trauma. This is because when the brain is injured, astrocytes go into overdrive to support the healing process, and they eventually form a scar that is visible on an MRI scan for a long time after the injury. This is analogous to how your skin grows scar tissue as it heals a cut.

Although we weren't expecting to see stroke-like changes in the hypothalami of people with obesity, we thought it was worth looking for a subtler version of the same scarring—similar to what we found in rats and mice. And that's exactly what we saw. Schur's analysis showed that the more signs of damage we found in a person's hypothalamus, the more likely he was to have obesity. What's more, this effect was once again located in the part of the hypothalamus that harbors NPY and POMC neurons. "The scariest implication," explains Schur, "is that the food we eat may cause damage in areas of the brain that we need to regulate our body weight and our appetite, as well as our blood sugar and, to some degree, our reproductive health."[121] Just as a tumor in Elisa Moser's hypothalamus caused her to develop obesity, a milder form of brain damage could be contributing to our own expanding waistlines.

Now it's time for a reality check: We still don't know to what extent these changes actually contribute to obesity, rather than being a result of obesity, or simply passively associating with the fattening process. We'll have to do more research to figure that out. However, what we can conclude is that the hypothalamus is under duress in obesity and that this is likely caused (at least in part) by the unhealthy food we eat. In response to this challenge, the hypothalamus activates a broad swath of cellular stress

---

121 Reproduction and blood glucose regulation are closely tied in with energy homeostasis in the brain. When a person (particularly a woman) doesn't have enough energy stores, the brain shuts down libido and reproductive function. The link between energy homeostasis and blood glucose regulation may be an important reason why obesity and diabetes often occur together.

response pathways, and some of these have the potential to dampen leptin signaling and contribute to the fattening process.[122] This likely operates in parallel with the set-point-altering effects of food reward and protein intake we encountered in the last chapter.

Brain damage is a daunting term, and it may make the situation seem hopeless for people who would like to lose weight. Yet our research also suggests that the process is reversible—at least in mice. When we switch mice off a fattening diet and back on to a strict healthy diet, even without restricting their calorie intake, they lose their excess fat, and their astrocytes and microglia go back to normal. This is true even if they've been obese for a long time. We still don't know if the same is true for humans, but there is reason to hope.

What's so fattening about the diets we use to make rodents obese in a research setting, and how do they injure the hypothalamus? In many ways, they are similar to the diets of affluent humans. They're made of refined ingredients; they have a high calorie density; they're highly rewarding (to rodents); they're high in fat and often sugar. The diet Schwartz and I used in our research is a light blue pellet that has the texture of greasy cookie dough.[123] Rodents greatly prefer this stuff to normal, unrefined, low-fat food, and when we give them unrestricted access to it, they gorge on it for the first week or so.

Many researchers have tried to narrow down the mechanisms by which this food causes changes in the hypothalamus and obesity, and they have come up with a number of hypotheses with varying amounts of evidence to support them. Some researchers believe the low fiber content of the diet precipitates inflammation and obesity by its adverse effects on bacterial populations in the gut (the gut microbiota). Others propose that saturated fat is behind the effect, and unsaturated fats like olive oil are less fattening. Still others believe the harmful effects of overeating itself, including the inflammation caused by excess fat and sugar in the bloodstream and in

---

122 Endoplasmic reticulum stress is another stress response pathway that has been implicated. Also, changes in cell turnover and the function of stem cells that are constantly replenishing neurons in the hypothalamus.

123 Research Diets D12492.

cells, may affect the hypothalamus and gradually increase the set point. In the end, these mechanisms could all be working together to promote obesity. We don't know all the details yet, but we do know that easy access to refined, calorie-dense, highly rewarding food leads to fat gain and insidious changes in the lipostat in a variety of species, including humans. This is particularly true when the diet offers a wide variety of sensory experiences, such as the hyperfattening "cafeteria diet" we encountered in chapter 1.

Personally, I believe overeating itself probably plays an important role in the process that increases the adiposity set point. In other words, repeated bouts of overeating don't just make us fat; they make our bodies want to *stay* fat. This is consistent with the simple observation that in the United States, most of our annual weight gain occurs during the six-week holiday feasting period between Thanksgiving and the new year, and that this extra weight tends to stick with us after the holidays are over. Thanksgiving dinner is the definition of overeating, and Christmas Eve, Christmas Day, and New Year's Eve aren't far behind. Throughout that entire period, well-meaning family and friends inundate us with cookies, pies, and other tempting calorie-rich treats that tend to hang around the kitchen until we eat them.

Because of some combination of food quantity and quality, holiday feasting ratchets up the adiposity set point of susceptible people a little bit each year, leading us to gradually accumulate and defend a substantial amount of fat. Since we also tend to gain weight at a slower rate during the rest of the year, intermittent periods of overeating outside of the holidays probably contribute as well.

How might this happen? We aren't entirely sure, but researchers, including Jeff Friedman, have a possible explanation: Excess leptin itself may contribute to leptin resistance. To understand how this works, I need to give you an additional piece of information: Leptin doesn't just correlate with body fat levels; it also responds to short-term changes in calorie intake. So if you overeat for a few days, your leptin level can increase substantially, even if your adiposity has scarcely changed (and after your calorie intake goes back to normal, so does your leptin). As an analogy for how this can cause leptin resistance, imagine listening to music that's too loud. At first, it's thunderous, but eventually, you damage your hearing, and the volume drops. Likewise, when we eat too much food over the course of a few days, leptin levels increase sharply, and this may begin to desensitize the brain circuits that respond to leptin. Yet Rudy Leibel's group has also shown that high leptin levels alone aren't enough—the hypothalamus seems to require a second "hit" for high leptin to increase the set point of the lipostat. This second hit could be the brain injury process we, and others, have identified in obese rodents and humans.

Let's recap what I've proposed thus far. We overeat because we're surrounded by seductive, calorie-dense food that's a great deal. The food's high reward value increases the set point of the lipostat, though not necessarily permanently, and this further facilitates overeating. At the same time, overeating itself spikes leptin levels and injures the hypothalamus by a mechanism we have yet to nail down (likely involving diet quality in addition to quantity). These two simultaneous hits cause the hypothalamus to lose sensitivity to the leptin hormone, meaning that it requires more leptin, and therefore more body fat, to hold off the starvation response that drives us to overeat. This time, the increase in your set point is permanent, or at least difficult to reverse. The lower limit of your comfortable weight creeps up.

To be clear, this is a working hypothesis that needs to be tested further before we can hang our hats on it. We don't yet have a complete understanding of how obesity develops and is maintained, but each year brings us closer to the answer.

EFFICIENTLY MANAGING OVEREATING

One practical implication of this research is that if you want to control your weight over the long term, focusing on the six-week holiday period will give you the greatest return on your effort. Developing strategies to avoid holiday overeating, such as getting rid of holiday snacks in the kitchen and cooking lighter versions of traditional recipes, might go a long way toward curbing the inexorable upward arc of adiposity that most of us experience over our lifetimes.

THE HUNGRY BRAIN

It's four in the afternoon, and I haven't eaten since breakfast. I also just rode my bicycle for over an hour to get to the University of Washington in Seattle. As I look at images of junk food inside an MRI machine in the mazelike basement of the Health Sciences Building, I can't help but think how surreal the experience is. I'm lying inside a Philips Achieva 3.0T, which looks like a giant white doughnut standing on its outer edge. My head, which is in the center of the doughnut hole, is secured by an inflatable cap, and I'm wearing earplugs to protect my ears from the deafening sound of the machine. I'm trying hard not to move.

Researchers and doctors use MRI to look for structural features of the brain, but they can also use it to measure brain activity.[124] This technique is called *functional MRI,* or fMRI for short. My colleague Ellen Schur and her team are using fMRI to understand the brain regions that determine hunger and satiety, ultimately influencing how much we eat.

In this particular experiment, I'm looking at blocks of images that fall into three categories: 1) rewarding, high-calorie foods, such as pastries, pizza, and potato chips; 2) low-calorie healthy foods, such as strawberries, celery, and apples; and 3) nonfood items like shoes and cars. While I look at the images, the MRI machine measures my brain activity using a magnetic field six hundred times stronger than a typical refrigerator magnet. By comparing my brain's response to the three categories of images, we can see which areas are activated by each.

The next week, I stop by Schur's office to go over the images. Schur and research scientist Susan Melhorn pull them up on a computer monitor. First, we look at an image in which they've compared my brain's response to high-calorie foods and my response to nonfood images. This allows us to subtract out the brain activity that happens when we look at pictures in general, and focus on changes that relate specifically to looking at high-calorie food.

"You have a classic response," notes Schur. First, she points out my VTA. If you recall from chapters 2 and 3, this is the brain region that sends dopamine into the ventral striatum. Both the VTA and the ventral striatum

---

124 fMRI measures changes in blood oxygen, which is a proxy for local neuron activity.

are central to motivation and reinforcement—for example, the desire we feel when we smell freshly baked brownies. "My VTA is lit up!" I exclaim, unable to contain myself. There's a bright, colorful blob where my VTA is located, indicating that my brain's dopamine system was very excited by the high-calorie food images. You can see a black-and-white version of the image in the upper left-hand panel of figure 37.

Next, we move on to my ventral striatum, which should also be activated since it receives input from the VTA. This time, the blob is even larger, as you can see in the upper right-hand panel of figure 37. "That's monstrous!" proclaims Schur.

"That's the biggest ventral striatum response I've ever seen," adds Melhorn.

They explain that skipping lunch and riding my bike to the medical center had created a much larger energy deficit than they typically see in their subjects, therefore provoking a heightened state of food motivation in my brain.

The next region we examine is my OFC, the place where economic value is computed during the decision-making process. Again, it's partially obscured by a colorful blob (upper right-hand panel of figure 37). "You need to make a decision," Schur offers by way of explanation, "and from here you're going to activate a plan to obtain the food."

The fourth region Schur examines is my insular cortex. This is a part of the brain that processes taste information. Again, it's covered by a colorful blob. I find this puzzling since I hadn't tasted the food, but Schur points out that the insular cortex is often activated by looking at food, similar to how our motor cortex fires when we think about movements. "We use some of the same neurons to think about movements that we use to make movements," explains Schur. Evidently, my brain was going through the motions of eating—rehearsing that glorious moment when I'd be stuffing pizza into my mouth (sadly for my long-suffering insular cortex, that never happened).

The diagnosis was clear: My hungry brain wanted food. A lot. And it wouldn't be satisfied by low-calorie foods, because the pattern of activation was almost nonexistent when I gazed at fruits and vegetables (see bottom two panels of figure 37). "When we're hungry," explains Schur, "our bodies don't want healthy food." Instead, powerful, instinctive brain regions draw

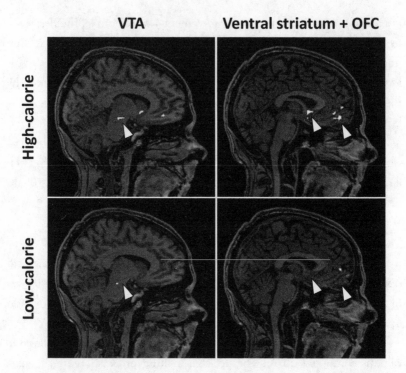

Figure 37. The author's brain response to high-calorie and low-calorie food images, superimposed on a static image of the author's brain. White areas indicate activation; arrows point to the VTA (left column), ventral striatum, and OFC (right column). Note the greater activation in response to high-calorie food images. All panels represent a ratio of the brain response to food images versus nonfood images. Irrelevant brain regions were masked to highlight areas of interest. Special thanks to Ellen A. Schur, Susan J. Melhorn, Mary K. Askren, and the University of Washington Diagnostic Imaging Sciences Center.

us toward concentrated, quick, easy calories. "That's what we're all up against."

My fMRI results are consistent with what Schur's studies have shown. When people are hungry, their brains react strongly to calorie-dense foods.[125] However, after they eat, this reaction to food cues subsides. Schur

---

125 You may be wondering: If the hypothalamus is involved in energy balance, why didn't it show up? We actually did see a signal in my hypothalamus, but it's hard to interpret for technical reasons that relate to the fMRI technique. Part of the problem is that energy homeostasis centers like the hypothalamus and brain stem are a collection of small, specialized nuclei, containing many cell types doing different things, and fMRI doesn't have the resolution to untangle this complexity.

explains: "At the end of a meal, what you're really experiencing is the fact that food doesn't look good to you anymore. The taste of it isn't as good as it used to be; you look at your plate and say, 'Ugh, I don't want any more of that.'" As a meal progresses, something in the brain receives information about what we've eaten and shuts down the circuits that make us want more food. How does this work, and can we exploit it to help curb our tendency to overeat?

WILL THE REAL SATIETY CENTER PLEASE STAND UP?

Harvey Grill, a neuroscientist at the University of Pennsylvania, studies the brain stem—a complex region of the brain where it joins the spinal cord. The brain stem is the most ancient part of the brain, evolutionarily speaking, and it tends to govern deeply instinctual, nonconscious functions, such as digestion, breathing, and basic movement patterns (see figure 38). It is also, according to Grill's research over the last forty years, the most important brain region for satiety.

"At the time I started my postdoc in 1974 at the Rockefeller Institute with Ralph Norgren," explains Grill, "there really wasn't a basis in data yet—there was only an idea." At the time, the prevailing view was that the hypothalamus is the only brain region that regulates food intake. Yet Grill and Norgren knew that the brain stem receives numerous inputs from the gut and mouth that are relevant to eating—and also contains outputs that regulate eating-related movements like chewing. "The question was," recalls Grill, "to what extent are these linked?"

Grill spent a year perfecting a technique that allowed his team to surgically inactivate everything in the rat brain except the brain stem and nearby structures. These "decerebrate" rats were unable to use most of the circuits in their brains, including their hypothalami. Surprisingly, when Grill's team placed food into the front of their mouths, they would chew and swallow normally. Even more impressive, when offered food continuously, decerebrate rats would eat the same amount as intact rats normally eat during a meal and then abruptly refuse additional food. "They would take meals!" exclaims Grill, still excited by his seminal finding four decades later.

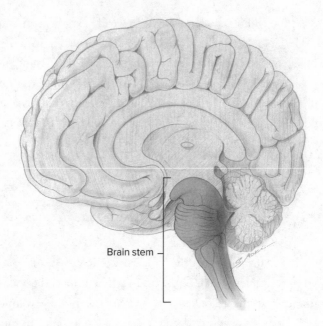

Brain stem —

Figure 38. The brain stem.

The similarity with normal rats didn't end there. Decerebrate rats reacted to a variety of satiety-related signals in the same way as normal rats: They ate less at a meal when Grill's team gave them a "snack" first, and they ate less in response to satiety hormones that the gut normally produces when we eat. This demonstrated, without a shadow of a doubt, that the brain stem is single-handedly capable of monitoring what's happening in the gut and generating the satiety response that ends a meal.[126]

Thanks to the research of Grill and many others, we now have a reasonably clear picture of how this happens. When you eat food, it enters your stomach and stretches it. After partially digesting the food, your stomach gradually releases it into the small intestine. Here, specialized cells in the

---

126 In the scientific literature, *satiation* is the sensation of fullness that eventually terminates a meal, while *satiety* is the state after a meal that makes you less likely to eat again. They are both characterized by a low motivational drive to eat, and they both result from similar mechanisms, so I'm not going to complicate the discussion by distinguishing between them in the main text. I'll use the term *satiety* to mean fullness and loss of interest in food, whether that occurs at the end of a meal or afterward. This is how the term is used colloquially.

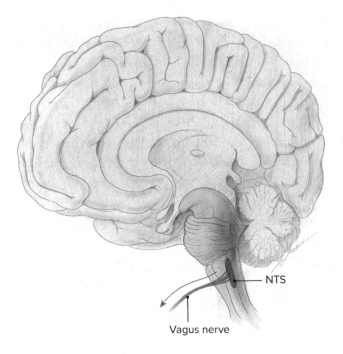

NTS

Vagus nerve

Figure 39. The nucleus tractus solitarius and the vagus nerve.

intestinal lining detect the nutrient content of what you ate, for example, the amount of carbohydrate, fat, and protein. These stretch and nutrient signals are relayed to the brain, primarily via the vagus nerve, which plays a major role in bidirectional gut-brain communication (see figure 39). At the same time, incoming nutrients cause the gut and pancreas to release a number of hormones that either activate the vagus nerve or act on the brain directly.

These signals that encode the quantity and quality of food you just ate converge on a brain region called the *nucleus tractus solitarius* (NTS), which is the junction point of the vagus nerve with the brain stem. The NTS integrates the various signals ascending from the digestive tract and produces a level of satiety that's appropriate for what you ate. These complex computations happen beyond your conscious awareness, and the only information your conscious brain receives is whether or not you feel full.

Despite the fact that part of the hypothalamus was once called the *satiety center,* and leptin was dubbed the *satiety factor,* we now believe the brain stem is the primary brain region that directly regulates meal-to-meal satiety, while leptin and the hypothalamus primarily regulate long-term energy balance and adiposity. Grill's research shows that although decerebrate rats take normal-sized meals, if they are underfed, they are unable to compensate normally by increasing the size of subsequent meals. In other words, their satiety system works great, but their lipostat is out of the picture, once again suggesting that the hypothalamus may be required for that function. A more accurate name for the hypothalamus would be the *adiposity center*, and leptin, the *adiposity factor*. Yet this distinction isn't black and white: Grill's research shows that the brain stem has a hand in regulating adiposity, and the hypothalamus may also play a role in regulating meal-to-meal food intake.

At some point, this information has to reach the action selection circuits that tell the brain whether or not to eat food. Although there are multiple connections from the hypothalamus and brain stem to the basal ganglia and associated structures, we still don't have a very clear picture of how they influence food-related decisions. However, Palmiter's and Lowell's research is increasingly converging on a small area of the brain stem called the parabrachial nucleus that receives input (direct and indirect) from NPY neurons, POMC neurons, and NTS neurons. The parabrachial nucleus could end up being a master regulator of hunger and satiety that plugs into action selection circuits—but this remains to be seen.[127]

Ultimately, the size of your meals has a major impact on your total calorie intake, and whether or not you gain weight over time. This is partially determined by how your brain stem generates satiety as you eat, eventually causing you to lose interest in food. Yet we know the hypothalamus impacts adiposity in large part by modifying food intake. How does this work? The hypothalamus influences brain stem satiety circuits in response to long-term changes in adiposity. In other words, if you're dieting and you've lost fat, the hypothalamus ensures that it takes more food to feel full at a meal than it did before you lost fat. Your brain dampens the feeling of satiety so you won't feel satisfied until you've eaten enough calo-

---

127 The parabrachial nucleus sends fibers to the intralaminar nucleus of the thalamus, which is the input nucleus for the basal ganglia in that region of the brain. This suggests that it could be an option generator. Alternatively, it could send information to the OFC and ventromedial prefrontal cortex, contributing to action selection by providing input about current energy status to economic choice circuits.

ries to start regaining fat. This is why people who are dieting often seem to have a bottomless appetite and never feel full. Conversely, if you've overeaten and gained fat, your brain enhances the feeling of satiety so your meals will be smaller for a while. This is how we think the hypothalamus and brain stem work together to regulate appetite and adiposity.

Since the hypothalamus influences brain stem satiety circuits, any impairment of the leptin system should increase the amount of food it takes a person to reach satiety. This is exactly what researchers have found in people with obesity, consistent with the idea that their brains are leptin resistant. When a person with obesity eats food, it doesn't suppress her brain's response to food cues as much as it does in a lean person. The brain regions that govern hunger and motivation just keep firing, driving her to overeat. This might seem grim, but there are ways to mitigate it.

The system of gut-brain communication that governs satiety doesn't do a perfect job of transmitting the calorie value of a meal to the brain. In other words, some foods make us feel more full than others, even if they contain the same number of calories. Because of this, we can exploit the quirks of the satiety system to help naturally reduce (or increase) our calorie intake, without discomfort.

In 1995, Susanna Holt and colleagues published a groundbreaking paper that gives us powerful insight into how we can use food to "trick" the brain into feeling full with fewer calories. The idea behind it is quite simple. Holt and her team recruited volunteers and fed them 240-Calorie portions of thirty-eight common foods, such as bread, oatmeal, beef, peanuts, candy, and grapes. Over the next two hours, the volunteers recorded how full they felt every fifteen minutes. Holt's team used the resulting data to calculate a "satiety index" for each food—representing how filling it is per calorie. Then, they analyzed the data set as a whole to determine which food properties are the most strongly related to satiety.

White bread, as expected, had a low satiety index relative to other foods, meaning it delivers little satiety per unit calorie. Whole-grain bread, in contrast, had a significantly higher satiety index. Calorie-dense bakery products, like cake, croissants, and doughnuts, had the lowest satiety index of all foods tested. Fruit, meat, and beans tended to have a high satiety index. Plain potatoes were off the charts—far more filling than any other food. Holt and colleagues noted that "simple 'whole' foods such as the fruits, potatoes, steak and fish were the most satiating of all foods tested."

Holt's team found that the sating ability of each item was largely explained by a few simple food properties. The first is calorie density; in other words, the volume of food per calorie.[128] For example, oatmeal is mostly water, so it has a much lower calorie density than crackers, which are nutritionally similar but contain little water. The lower the calorie density of the item, the more satiety it produced per unit calorie—and the effect was extremely robust. This makes sense, because stomach distension is one of the main signals that the NTS monitors to regulate satiety. If your stomach contains more food volume, you'll feel more full, even if that food doesn't contain more calories. This only works up to a point, however; you can only trick the brain so much. A belly full of lettuce isn't going to cut it.

Neck and neck with calorie density was another factor we've encountered before: palatability. The more palatable a food, the less filling it was. Again, this makes sense. Palatable foods are those that the brain intuitively views as highly valuable, and the brain is quite good at removing barriers to their consumption. We even have ideas about how this might work. Within the hypothalamus lies a region called the lateral hypothalamus (LH), which is a nexus between energy balance and food reward functions (among other things). Researchers have known for a long time that stimulating the LH causes animals to eat voraciously, and disrupting it makes them lean. As it turns out, palatable food activates neurons in the LH. Furthermore, the LH sends fibers directly to the NTS of the brain stem, where it inhibits neurons that play a role in satiety—so it's not much of a leap to suppose that eating palatable foods might inhibit the very NTS neurons that make us feel full. This may be one of the reasons why we tend to overeat highly palatable foods and magically grow a "second stomach" for dessert. Sticking with simple foods can help us restrain our calorie intake without feeling hungry.

The third most influential factor Holt and colleagues identified is a food's fat content. The more fat it contained, the less filling it was per calo-

---

128 Technically, calorie density usually refers to mass rather than volume per calorie, but volume is more intuitive, and it's the property your stomach measures.

rie. People often find this counterintuitive, because when they eat high-fat foods, they feel extremely full. The key to understanding this is to remember that we're talking about fullness *per unit calorie*. If you eat a stick of butter, you may feel full, but you will also have eaten over 800 Calories—the equivalent of two and a half large baked potatoes. Isolated fats like butter and oil are the most calorie-dense substances in the human diet by far, mostly because fat delivers nine Calories per gram versus only four for carbohydrate and protein. Isolated fats also increase the palatability of food. For these reasons, adding fat to food is a highly effective way to increase your calorie intake without increasing your satiety much, and limiting added fat helps reduce calorie intake without sacrificing satiety.

That being said, fat in general isn't necessarily something you need to avoid if you want to control your calorie intake. Research has shown that the reason fat makes us eat more is precisely because of its high calorie density and palatability. When high-fat foods aren't calorie dense or highly palatable, they provide the same level of satiety per calorie as high-carbohydrate foods. What this means is that if we eat fat in the context of unrefined, filling foods like meat, fish, eggs, dairy, nuts, and avocados, a higher fat intake can be compatible with a naturally slimming diet pattern. While these foods are high in fat, they don't have the deadly combination of calorie density and extreme palatability that characterize other high-fat foods like potato chips and cookies.

A fourth critical factor that Holt's team identified is fiber. The more fiber a food contained, the more filling it was. This explains why whole-grain bread is more filling than white bread, despite their similar calorie density.

Finally, the protein content of a food was a major contributor to satiety. This is consistent with a large body of research showing that protein is more filling than carbohydrate or fat, per unit calorie. Both the lining of the small intestine and the pancreas have the ability to detect dietary protein, and they relay this signal to the NTS. For reasons that aren't entirely clear, this protein signal seems to play a disproportionate role in satiety. Coupled with the effects of protein on the lipostat we discussed in the last chapter, this may explain why high-protein diets help people eat less and lose fat without feeling hungry.

Holt's results go a long way toward explaining why both rats and humans overeat spectacularly on a cafeteria diet, and why we overeat without intending to in our daily lives. The nonconscious parts of the brain that regulate satiety, including the NTS, respond to specific food properties, such as food volume, protein, fiber, and palatability. Many of our modern processed foods have properties that don't stimulate satiety circuits to the same degree as traditional whole foods. These foods, such as pizza, ice cream, cake, soda, and potato chips, invariably boast a combination of properties that make them less filling per calorie. Since most people use the sensation of satiety as a signal to stop eating, these foods allow us to blow past the point where we've had enough to satisfy our calorie needs—yet we don't even realize we're overeating because we don't feel any fuller at the end of the meal.

Traditional diets from many different cultures tend to have the opposite qualities: lower calorie density and palatability, and higher fiber (although not necessarily higher protein). The most striking antithesis of modern junk food, however, is the popular Paleolithic diet, inspired by the diets of our hunter-gatherer ancestors. This whole-food-based diet combines multiple sating food properties, including high protein, high fiber, low calorie density, and moderate palatability. In contrast to the caricature that most people are familiar with, the Paleolithic diet isn't an all-meat diet (hunter-gatherers didn't eat bacon, sorry), isn't necessarily low in carbohydrate, and contains large amounts of whole plant foods. Research confirms that it's highly sating and naturally promotes a lower calorie intake. In clinical trials, the Paleolithic diet has often outperformed conventional diets both for weight loss and metabolic improvements, explaining its popularity.

At the end of the paper, Holt and her colleagues put the pieces together for us: "The results therefore suggest that 'modern' Western diets which are based on highly palatable, low-fibre convenience foods are likely to be much less satiating than the diets of the past or those of less developed countries." Fortunately, Holt's findings also empower us to do something about it.

Yet not everyone reacts the same way to the modern food environment. Some people don't overeat in the first place, most of us tend to overeat and gain weight, and a few lucky folks don't gain weight *despite* eating prodigious amounts of food. What explains these differences?

EATING FOR SATIETY

If your goal is to eat fewer calories without going hungry, high-satiety foods will help. These are items that have some combination of low calorie density, moderate palatability, high protein, and/or high fiber, such as beans, lentils, fresh fruit, vegetables, potatoes and sweet potatoes, fresh meat and seafood, oatmeal, avocados, yogurt, and eggs. The exceptionally high satiety value of potatoes is undoubtedly one of the reasons why the "potato diet" we encountered in chapter 3 is effective. But if you cover your potato with calorie-dense flavorings such as butter and cheese, or deep-fry it into french fries, all bets are off.

BORN TO BE FAT

In 1976, Mats Börjeson, a researcher at Lund University, Sweden, published a paper that rocked the research community—and whose implications continue to challenge many of our dearly held notions about obesity. Börjeson wanted to understand the importance of genes in determining who develops obesity and who doesn't—a question that had not yet been seriously investigated. To answer this question, he exploited a neat trick of biology: the fact that identical twins are genetically identical, while nonidentical (fraternal) twins share only half their genetic material. In both cases, the siblings develop in the same womb, are born at the same time, and are raised in the same family, so the only significant difference between identical and fraternal twins is their degree of genetic relatedness.

Researchers can exploit this principle to understand what proportion of a trait is genetically determined. For example, if identical twins have skin colors that are more similar than fraternal twins, it implies that skin color is genetically influenced. This makes sense: The less two people are related, the less similar they tend to be. As common sense suggests, skin color is heavily influenced by genetics.

Börjeson recruited forty identical and sixty-one fraternal twins and measured their body weights. He found that identical twins tended to have very similar body weights, while fraternal twins were more divergent. "Genetic factors," he concluded, "apparently play a decisive role in the origin of obesity."

Since Börjeson's study, many others have confirmed that genes have an outsized influence on adiposity. In fact, in modern affluent nations like the United States, genetic differences account for about 70 percent of the difference in body weight between individuals. They also play a prominent role in many of the details of our eating behavior, such as how much food we eat at a sitting, how responsive we are to the sensation of fullness, and how much impact food reward has on our food intake. In other words, whether a person is lean or fat in today's world has less to do with willpower and gluttony and more to do with genetic roulette.[129] If you want to be lean, the most effective strategy is to choose your parents wisely.

Genes also explain that friend of yours who seems to eat a lot of food, never exercises, and yet remains lean. Claude Bouchard, a genetics researcher at the Pennington Biomedical Research Center in Baton Rouge, Louisiana, has shown that some people are intrinsically resistant to gaining weight even when they overeat, and that this trait is genetically influenced. Bouchard's team recruited twelve pairs of identical twins and overfed each person by 1,000 Calories per day above his calorie needs, for one hundred days. In other words, each person overate the same food by the same amount, under controlled conditions, for the duration of the study.

If overeating affects everyone the same, then they should all have gained the same amount of weight. Yet Bouchard observed that weight gain ranged from nine to twenty-nine pounds! Identical twins tended to gain the same amount of weight and fat as each other, while unrelated subjects had more divergent responses. Furthermore, not only did twins gain a similar amount of fat, they even gained it in the same places. If one person stored the excess fat inside his abdominal cavity—the most dangerous place to gain fat—his twin usually did the same. Not only do some people have more of a tendency to overeat than others, but some people are intrinsically more resistant to gaining fat even if they do overeat.

---

129 Although cognitive traits like willpower and the propensity to overindulge are probably genetically influenced as well.

The research of James Levine, an endocrinologist who works with the Mayo Clinic and Arizona State University in Scottsdale, Arizona, explains this puzzling phenomenon. In a carefully controlled overfeeding study, his team showed that the primary reason some people readily burn off excess calories is that they ramp up a form of calorie burning called "non-exercise activity thermogenesis" (NEAT). NEAT is basically a fancy term for fidgeting. When certain people overeat, their brains boost calorie expenditure by making them fidget, change posture frequently, and make other small movements throughout the day. It's an involuntary process, and Levine's data show that it can incinerate nearly 700 Calories per day! The "most gifted" of Levine's subjects gained less than a pound of body fat from eating 1,000 extra Calories per day for eight weeks. Yet the strength of the response was highly variable, and the "least gifted" of Levine's subjects didn't increase NEAT at all, shunting all the excess calories into fat tissue and gaining over nine pounds of body fat. For Levine, the study highlighted the importance of light physical activity throughout the day, which inspired him to invent the treadmill desk.[130]

Together, these studies offer indisputable evidence that genetics plays a central role in obesity and dispatch the idea that obesity is primarily due to acquired psychological traits. Yet they don't tell us which genes are responsible for individual differences in eating behavior and obesity. For that, researchers would need different approaches.

Stephen O'Rahilly and Sadaf Farooqi haven't been twiddling their thumbs since they identified humans who lack leptin; in the meantime, they've located a number of other single-gene mutations that cause severe obesity. As it turns out, nearly all of them are in the leptin signaling pathway. These genes disrupt either leptin itself, the leptin receptor, or downstream signals that mediate leptin's actions in the brain. What's more, now that researchers have identified the genes responsible for fattening our various rodent models of single-gene obesity, they've found that these genes are in the leptin signaling pathway as well. It's not every day in science that results from several related fields of research converge in such a powerful

---

130 As ridiculous as it sounds and looks, the treadmill desk is probably a really good idea from a health and weight standpoint. I look forward to more studies on its use in the workplace.

way. In this case, they all say in chorus: Leptin signaling in the brain is a key component of the biological control of adiposity.

O'Rahilly and Farooqi focus on children with extreme obesity. These are the cases where something really seems to be broken, and it isn't so farfetched to suppose that a severe genetic disruption could be responsible. Among these patients, about 7 percent show evidence of such genetic disruptions. The most common mutation disrupts the melanocortin-4 receptor, which is primarily responsible for the appetite-suppressing effects of melanocortins in the brain (the substances released by POMC neurons). In these children, melanocortin release triggered by their own circulating leptin fails to restrain their appetites, so they eat substantially more than typical children. However, according to Farooqi, known mutations only account for about 1 percent of adults admitted to an obesity clinic. They don't explain the obesity epidemic, yet, Farooqi adds, "we think there's an awful lot more to find."

Most researchers think destructive single-gene mutations only account for a small fraction of overeating and obesity. Yet at the same time, we know that genetics explains much of the differences in eating behavior, and most of the differences in adiposity, between individuals. Where are the missing genes? To answer this question, we have to turn to another type of genetics research that studies common gene variation rather than rare, catastrophic mutations that disrupt entire signaling pathways.

Many genes come in several different flavors, called *alleles,* each of which is encoded by a slightly different DNA sequence. Sometimes, all alleles of a particular gene are functionally identical, but in other cases, they have subtly different functions. For example, different alleles of the same set of genes determine our eye colors and blood types. This common form of genetic variation explains much of why individual humans look and act differently from one another.

Researchers have figured out ways to determine which genes have different alleles that lead to different body weights. Unlike O'Rahilly and Farooqi's studies, which focus on selected cases of extreme obesity, these methods are aimed at identifying the genetic factors that determine adiposity in the general population.

Remarkably, it turns out that a boatload of genes influence adiposity in humans, but each individual gene only has a small impact. So far, researchers have identified nearly one hundred genes that influence adiposity—yet together, these genes explain less than 3 percent of the differences in adiposity between people. Clearly, there's still work to do. However, by looking at the genes that have been identified so far (which were identified precisely because they're the most influential), we can gain unbiased insights into the biological processes that underlie garden-variety obesity.[131] By now, you should have a pretty good guess what organ these genes tend to influence: the brain. Although some genes relate to other processes such as fat metabolism, the lion's share exert their influence in the brain, and of those, many act via brain circuits that are already known to regulate food intake and adiposity (such as POMC neurons and their downstream targets). This suggests that genetic differences in brain function are the primary reason why some people are fatter than others.

Well, I suppose that's that: If you have fat genes, you're destined to be fat. Or are you? A century ago in the United States, people carried the same genes we do today, yet few people had obesity. What has changed isn't our genes, it's our *environment*—our food, our cars, our jobs. This leads us to a critical conclusion about obesity genes: In most cases, they don't actually make us fat, they simply make us susceptible to a fattening environment. In the absence of a fattening environment, they rarely cause obesity. As Francis Collins, geneticist and director of the National Institutes of Health, is fond of saying, "Genetics loads the gun, and environment pulls the trigger." Unless you have a faulty gun, which is rare, if you don't pull the trigger, it doesn't discharge.

A few lucky people are so genetically resistant to obesity that they're unlikely to develop it under any circumstances. A few others are so genetically susceptible that they may carry excess fat even in a very healthy environment. For the rest us, the environment in which we find ourselves has a major impact on our weight.

---

131 Unbiased because the genes were identified as associating with adiposity without any prior knowledge or assumptions about the genes' locations or functions.

8

RHYTHMS

At the New York Obesity Research Center of Columbia University, assistant professor Marie-Pierre St-Onge randomly assigned her research volunteers to one of two groups. For five consecutive nights, the first group was granted nine hours in bed per night, while the second group was condemned to only four (unfortunately for the former group, they had to swap interventions in the second phase of the study[132]). Both groups slept and ate in the lab during that time so that researchers could closely monitor them. Each night, the volunteers were outfitted with a tangle of electrodes and wires that monitored brain waves and other indicators of sleep.[133]

The goal of this intensive study was to understand the impact of sleep restriction on food intake and brain activity, a topic that has captured the attention of many researchers in recent years—and which may help explain why we overeat in the modern world.

On the fifth day of St-Onge's study, the volunteers were turned loose upon the world and allowed to eat whatever they wanted for a day. The

---

132 This is called a *crossover* design. It's a powerful way to do studies because each person is compared against himself, reducing variability in measurements.

133 This is called polysomnography.

only catch was that they had to let the research team weigh and record every-thing they chose. At the end of the study, St-Onge and her colleagues ana-lyzed the data and came to a striking conclusion: Their volunteers ate nearly 300 more Calories per day when they were sleep-deprived than when they were well-rested. "In our experience," explains St-Onge, "sleep restriction increases food intake. It's as simple as that."

DYING TO SLEEP

To understand how sleep restriction causes us to overeat—and what we can do about it—we first need to understand the biological basis of sleep. The story begins in 1916, when the Viennese neurologist Constantin von Economo began seeing patients with a previously unknown brain dis-order. People afflicted with the disorder would sleep excessively—up to twenty hours per day—scarcely leaving time for other activities. *Encepha-litis lethargica,* as von Economo named the disorder, swept through Europe and North America in the early twentieth century, afflicting as many as one million people. Most patients were left either dead or permanently dis-abled by the brain damage it caused. By 1928, the disease vanished as sud-denly as it had appeared, and few cases have been reported since. To this day, we still aren't sure what caused it, although many believe it was trig-gered by an infectious agent.

Naturally, von Economo was interested in the biological basis of en-cephalitis lethargica, so he performed a series of brain autopsies on patients who had suffered from the disease. What he found was remarkable: patients with encephalitis lethargica invariably had brain damage at the junction between the upper brain stem and the lower forebrain, as illustrated in figure 40.[134] These findings led him to propose that the region affected by encephalitis lethargica contains an *arousal system* that normally keeps the rest of the brain awake.

Over the course of the ensuing century, our understanding of sleep has grown considerably, yet von Economo's observations have withstood the

---

134 A small subset of patients actually suffered from profound insomnia, and they had brain dam-age in a nearby, but clearly different, brain area—the ventrolateral preoptic area.

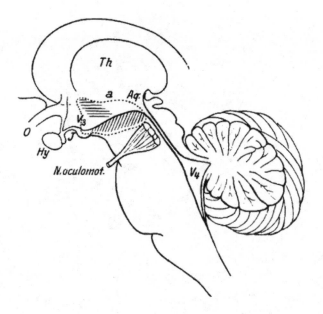

Figure 40. Brain lesions associated with encephalitis lethargica. From a 1926 paper by von Economo. Reprinted in Triarhou et al., *Brain Research Bulletin* 69 (2006): 244. The brain region of interest is bounded by a dotted line. Within the dotted line, diagonal hatching marks the location of lesions associated with excessive sleeping.

test of time. The brain does contain an arousal system that creates wakefulness and alertness, and part of this system lies precisely in the area he identified.[135] We now know that the arousal system has multiple component brain regions, most of which are located in the brain stem and hypothalamus (see figure 41). These regions send a broad network of fibers throughout much of the brain, releasing chemicals like dopamine, serotonin, norepinephrine, and acetylcholine, which keep us awake and alert.

The brain also contains a complementary *sleep center,* located in a part of the hypothalamus called the *ventrolateral preoptic area* (VLPO). When it's time to sleep, the VLPO sends signals to the various parts of the arousal system, shutting them down and allowing the brain to disengage from the outside world (see figure 42).

Cliff Saper, a sleep neuroscientist at Harvard University, has shown that

---

135 Encephalitis lethargica primarily killed groups of cells that produce dopamine, which—as we've seen—plays a role in energizing our behavior (as opposed to groggy lethargy).

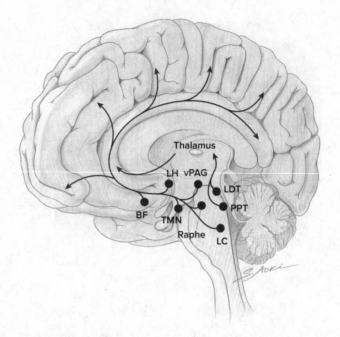

Figure 41. The arousal system. Nuclei in the brain stem and hypothalamus send projections to many brain areas, increasing arousal. BF, basal forebrain; LH, lateral hypothalamus; TMN, tuberomammillary nucleus; vPAG, ventral periaqueductal gray; LDT, laterodorsal tegmentum; PPT, pedunculopontine tegmentum; LC, locus coeruleus. Adapted from Saper et al., *Nature* 437 (2005): 1257.

the sleep center and the arousal system are the yin and yang of sleep and wakefulness. Each one inhibits the other, such that when one is active, the other is shut down. Saper's former graduate student Thomas Chow, who happens to have an electrical engineering degree from the Massachusetts Institute of Technology, pointed out to Saper that engineers refer to this arrangement as a "flip-flop switch." This is the kind of circuit you design when you want a system to fully occupy one of two states, such as wakefulness or sleep, but nothing in between. "It's Electrical Engineering 101," explains Saper. In our case, this flip-flop switch means we're either awake or asleep, without much middle ground.

Since sleep and wakefulness are stable states, there has to be a signal that's strong enough to kick one state into the other, or else we'd never fall asleep or wake up. We know that when we sleep too little, do hard physical or

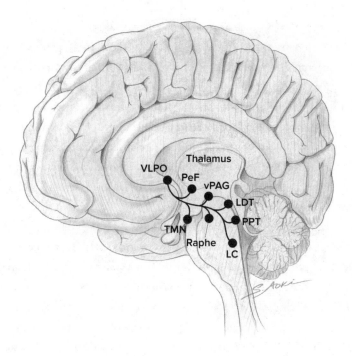

Figure 42. The ventrolateral preoptic nucleus, the brain's "sleep center." The VLPO sends projections to many areas of the arousal system, shutting them down and promoting sleep. TMN, tuberomammillary nucleus; PeF, perifornical area; vPAG, ventral periaqueductal gray; LDT, laterodorsal tegmentum; PPT, pedunculopontine tegmentum; LC, locus coeruleus. Adapted from Saper et al., *Nature* 437 (2005):1257.

mental work, or stay up longer than usual, we feel sleepy and we're more likely to doze off. This suggests that some sort of sleep-inducing signal accumulates in the brain, and the longer we're awake, the harder we work, the more it builds up. We now have strong evidence that a chemical called *adenosine* is that signal. Adenosine builds up in the brain while we're awake, and it builds up even faster when we exert ourselves. As it accumulates, it begins to inhibit the arousal system and activate the VLPO. Eventually, when adenosine builds up sufficiently, it triggers the flip-flop switch, and we fall asleep. During sleep, the brain clears excess adenosine, restoring our wakefulness by morning. Caffeine works by blocking adenosine's actions.

Common sense suggests that sleep is an important restorative process for the brain and the body, and research is increasingly backing this up.

During sleep, the brain clears waste products that accumulate during the day as a result of normal metabolism, including adenosine. It gently remodels itself, reinforcing important connections and pruning unimportant ones. And it mops up the protein amyloid-β, which is implicated in the development of Alzheimer's disease, one of the most tragic scourges of aging.

The fundamental importance of sleep is highlighted by the fact that all animals with a nervous system sleep, or at least enter sleeplike states. Saper explains that "every species that we've looked at, even going back to simple invertebrates like sea slugs, tend to have rest-activity cycles." He believes animals need sleep because neurons require periods of rest for the biochemical processes that support learning.

The restorative processes that happen during sleep are extremely important for the optimal functioning of the brain, and interfering with them leads inevitably to degraded performance of numerous brain functions—and worse. David Dinges, a sleep researcher at the University of Pennsylvania, has spent much of his career studying the effects of sleep restriction on brain function. In one particularly telling study, Dinges's team assigned volunteers to eight hours, six hours, or four hours in bed per night, for two weeks. Every two hours of waking time throughout the study, volunteers had to complete a battery of tests designed to measure various aspects of cognitive performance, including reaction time, attention, working memory, and basic arithmetic abilities. Their conclusion should serve as a caution to people who think sleep is a waste of time: "Chronic restriction of sleep periods to 4 h or 6 h per night over 14 consecutive days resulted in significant cumulative, dose-dependent deficits in cognitive performance on all tasks."

In other words, by a variety of measures, people who slept four or six hours per night showed substantially worse cognitive performance than people who slept eight hours per night. What's more, these deficits grew with each additional night of short sleep. Importantly, volunteers felt sleepier for the first few days of sleep restriction, but after that, "subjects were largely unaware of these increasing cognitive deficits." This suggests that people who short themselves on sleep may not even be aware of how poorly they're performing.

In the extreme, sleep loss can be deadly. This is illustrated by a rare genetic disease called familial fatal insomnia. As the result of a neurodegenerative process, patients with fatal familial insomnia gradually lose the ability to sleep, descending into hallucinations, delirium, dementia, and after a period of months, death. Researchers aren't sure to what extent death results from sleep loss rather than other aspects of the neurodegenerative process, but their condition does deteriorate rapidly after they develop insomnia. Older experiments in rats also back up the hypothesis that chronic sleep deprivation can kill. Yet even when sleep loss doesn't kill us, insufficient or poor-quality sleep can have insidious effects on brain activity and metabolic processes that tend to favor weight gain.

SWEET DREAMS

Returning to St-Onge's experiment from the beginning of the chapter, her team didn't stop at measuring food intake in her sleep-restricted volunteers. They also performed fMRI to see how sleep restriction affects the brain's response to food cues. The results of these brain scans suggest that sleep restriction increases the brain's responsiveness to food—particularly calorie-dense junk food like pizza and doughnuts. The parts of the brain associated with food reward, including the ventral striatum, were more active in sleep-restricted volunteers, perhaps explaining why they ate more.

Interestingly, the pattern of brain activity St-Onge observed in her sleep-restricted volunteers is quite similar to the pattern Rudy Leibel observed in the brains of people who had previously lost weight—and also the pattern Ellen Schur observed in my own brain when I was looking at food images in a hungry state. This suggests that lack of sleep doesn't just impair our cognitive functions; it may also impair the lipostat that senses the body's energy status and sets our motivation for food. "The brain is basically telling you that you're in a food-deprived state when you really are not," explains St-Onge. "It primes you for overeating and trying to overcome negative energy balance that's really not there." Essentially, when you don't sleep enough, your lipostat mistakenly thinks you need more energy, which activates your food reward system and causes you to eat more without intending to and often without even realizing it.

An aside: Astute readers may note that when you sleep less, it also increases your calorie expenditure, because your metabolic rate is higher during wakefulness than during sleep. St-Onge and her team have shown that limiting sleep to four hours a night does indeed increase calorie expenditure, but only by about 100 Calories per day. Since her sleep-deprived volunteers ate nearly 300 excess Calories daily, that still leaves 200 Calories to accumulate around their midsections at the end of the day—enough to turn a susceptible person overweight over time.

Time for another reality check: Although many studies have shown that sleep restriction increases the brain's response to food, increases calorie intake, and sometimes increases body weight, these studies haven't lasted longer than two weeks. To get an idea of whether this effect persists over the long term and results in gradual weight gain, we must turn to long-term observational studies that measure habitual sleep duration and see how it correlates with weight changes over time. At this point, researchers have conducted many such studies, and they overwhelmingly show that adults who sleep six or fewer hours per night do tend to gain more weight over time than those who sleep seven to nine hours per night.[136] It's also worth noting that the association between short sleep and weight gain is particularly strong in children. Although these studies by themselves can't establish a cause-and-effect relationship between sleep duration and weight gain, when considered alongside the controlled studies we've just discussed, they make quite a compelling case that part of the reason why we eat too much and gain weight is that we don't sleep enough.

This conclusion has disturbing implications for those of us who live in affluent, modern societies, because many of us don't get enough sleep. Twenty-nine percent of American adults get six hours of sleep or fewer per night, up from 22 percent in 1985. Researchers have reported similar trends among adolescents. Short sleepers are not only more likely to have obesity than people who sleep more; they're also more likely to develop chronic diseases, such as cardiovascular disease and diabetes, and more likely to die overall. Contrary to the claims we often encounter in the media and even sometimes in the scientific literature, our average time in bed in the United

---

136 Interestingly, long sleepers (more than nine hours) also tend to gain more weight over time in some studies. Certain researchers have even gone so far as to suggest that we should all sleep seven to eight hours a day, even if that isn't enough to feel rested. However, many people, including myself, are skeptical that sleeping for more than nine hours actually causes weight gain. It may simply be the case that people who sleep longer often do so because of depression or certain medications, both of which can favor weight gain.

HOW MUCH SLEEP DO YOU NEED?

Different people need different amounts of sleep to feel rested. Although this remains to be tested scientifically, it's possible that naturally short sleepers get the same amount of benefit in, say, six hours, that long sleepers get in nine hours. If true, this would suggest that what harms us isn't necessarily short sleep itself, but rather sleeping less than each of us needs as an individual. For example, a person who sleeps six hours per night but feels fully rested when she wakes up may not have anything to gain from attempting to sleep more, while another person who sleeps six hours per night but wakes up feeling miserable may benefit from more shut-eye. For now, the best advice is probably to sleep as much as you need to feel fully restored, rather than following rigid sleep guidelines that are based on averages from population studies.

States hasn't changed drastically over the last two decades, but it has declined slightly. I suspect this small decline is due to the meteoric rise of mentally stimulating digital media, such as video games and the Internet. We also have access to bright electric light at night and stimulating media at any time of day. Compare that to our distant ancestors, whose only source of light at night was a campfire or perhaps a candle and whose only source of evening entertainment was one another.

Although our average time in bed may not have changed much over the years, the amount of time in bed we actually spend getting high-quality sleep has dropped more substantially due to the sharp increase in the prevalence of sleep-disordered breathing conditions such as sleep apnea. Researchers attribute this problem to our expanding waistlines over the last forty years, which increases the volume of soft tissues around the airway that can interfere with breathing at night. This may create a vicious cycle in which fat gain leads to sleep-disordered breathing, which leads to sleep loss, which further exacerbates overeating and fat gain.

Yet sleep loss doesn't just affect the lipostat—it also undermines our ability to control the very impulses it activates.

A PRISONER OF YOUR IMPULSES

Sleep loss also favors overeating by affecting how we perceive risks and rewards. In 2011, Duke sleep researcher Michael Chee published a paper suggesting that sleep loss has far-reaching impacts on our economic decision-making behavior. Chee and his team had twenty-nine young adult volunteers either sleep normally or skip one night of sleep, and then he asked them to perform a series of experimental gambling sessions while lying in an fMRI machine.

Chee's results showed that pulling an all-nighter causes people to become less concerned about potential losses and more attracted to potential gains—basically, they become risk takers. Furthermore, this effect correlated with measurable differences in brain activity. People who hadn't slept showed greater activation of reward-related regions of the brain such as the ventral striatum in response to gambling gains, while also showing diminished brain responses to losses.

"Generally, with sleep loss you have a shifting of your economic preferences," explains Dan Pardi, a graduate student in the lab of Stanford sleep researcher Jamie Zeitzer. Researchers call this effect an *optimism bias,* and Pardi wondered if it might also apply to eating behavior. As we discussed in chapter 5, the brain weighs the potential costs and benefits of eating as it decides what and how much to eat. To revisit an earlier example, the brain may have to decide between keeping three dollars in your wallet and putting a pastry in your belly. Often, when we're faced with unhealthy food, the benefit is the immediate reward we get from eating something tasty, while the cost is the long-term impact on our adiposity and health. So Pardi hypothesized that if he added insufficient sleep to the mix—and consequently, optimism bias—people might also be more attuned to the benefits of eating than the costs. And this might nudge them toward unhealthy food choices.

To test his hypothesis, Pardi's team recruited fifty volunteers who believed they had signed up for a study on how sleep loss affects cognitive function. They were divided into seven groups, and each group was told to spend a different amount of time in bed the night preceding the experiment. These prescribed sleep times ranged from 60 to 130 percent of each person's usual time in bed. (A key advantage of Pardi's study is that it modeled a realistic range of sleep times that we commonly see in the general population, as opposed to many other sleep restriction studies, which allowed subjects as little as four hours of sleep, or none at all.) Seven days before the change in sleep time, Pardi's team gave each person a battery of tests to assess their typical level of alertness and feelings of sleepiness. On the day following the change in sleep time, they repeated the same tests.

Yet Pardi had a trick up his sleeve, on which the whole experiment hinged. While the volunteers took a break between tests, he had them watch two forty-minute movies that were unrelated to food. And during that time, Pardi put out bowls of foods like gummy bears, toffee peanuts, apple rings, and almonds that the volunteers could graze on. Pardi's team covertly measured each person's food intake by weighing the bowls before and after the movie break. At the end of the study, volunteers filled out a questionnaire rating how much they liked each food and how healthy they thought each item was.

Consistent with St-Onge's findings, sleepier participants munched more calories during the movie break. Yet they weren't just more likely to eat food in general—they were specifically more likely to eat food they themselves rated as both delicious and unhealthy. Just as Chee's results had predicted, Pardi's sleep-deprived volunteers seemed to be more compelled by the immediate reward of eating tempting foods than by the long-term costs. "When you have inadequate sleep," explains Pardi, "you're probably less likely to live in accordance with your own health goals. You're less likely to get into bed on time, you're less likely to go to the gym, and you're less likely to have your eating behaviors align with your long-term health goals." fMRI research supports Pardi's interpretation, showing that one night of total sleep deprivation reduces the food cue responsiveness of brain regions that support planning, reasoning, and long-term goals, while at the same time enhancing the responsiveness of brain regions that drive food reward. Together, this suggests that when you don't sleep enough, you're a prisoner of your own impulses—and those impulses will tell most of us to overeat unhealthy food.

But we know that not all sleep is the same. Why is sleep more restorative when it happens at the time we're used to sleeping? To answer this question, we'll have to return to the brain, exploring a system that may be able to make us fatter without even increasing our calorie intake.

THE DARKEST NIGHT

So far, we've covered some of the key brain regions that regulate sleep and wakefulness, ultimately impacting eating behavior—yet there's more to the story. Why do we feel compelled to sleep at night and not during the day? Why do we feel awful and perform poorly if we have to undertake a task when we would normally be sleeping?

In 1962, a French geologist and cave explorer named Michel Siffre conducted an unusual experiment that paved the way to answering these questions. It was the height of the Space Race, and scientists were eager to understand how the human body and mind would react to the absence of time cues that we might experience during space travel.

To find out, Siffre lowered himself into the Abyss of Scarasson, an inky black cave below the surface of the French-Italian Maritime Alps. He made his camp on a small subterranean glacier 427 feet below the surface, where the temperature hovered just below freezing and the relative humidity was 98 percent. There he stayed, cold, wet, and alone, for sixty-three days.

Siffre didn't bring a watch or anything else that could possibly reveal what time of day it was. He phoned his research colleagues on the surface to indicate each time he went to sleep, awoke, and ate.

When Siffre emerged from the Abyss of Scarasson and analyzed his results, he realized something remarkable: the length of his wake-sleep cycle had remained close to twenty-four hours for his entire two-month *séjour* in the cave. Yet because his cycle was slightly longer than twenty-four hours, it gradually desynchronized from the day-night cycle of the sun.

This suggested that the human body must quite literally contain a twenty-four-hour(ish) clock. Research over the ensuing half century has confirmed this, although we now know there is more than one clock. In fact, there are thirty-seven trillion of them: a tiny molecular timepiece in nearly every one of your cells. Together, these clocks synchronize many of the body's functions with the twenty-four-hour cycle of the sun—creating what is called a *circadian rhythm*. Our sleeping and waking cycle tends to follow a circadian rhythm, and so do our cognitive performance, eating behaviors, digestive functions, metabolic processes, and many other aspects of our behavior and physiology. If you've ever experienced the brain fog, lack of hunger, and digestive discomfort of jet lag, you understand the importance of circadian rhythms in regulating cognition, eating behavior, and digestion.

Your body's thirty-seven trillion cellular clocks are all kept more or less in synchrony by a master clock residing in a part of the hypothalamus called the suprachiasmatic nucleus (SCN). If we think of the cellular clocks in the body as a giant orchestra, the SCN is the conductor.[137] The master clock, in turn, takes its cues from the retina, the light-sensitive film of cells in the back of the eye, which detects the day-night cycle of the sun (see figure 43). But due to the characteristics of the retinal cells that transmit this information, the SCN only responds to blue light, which happens to be most abundant at midday.

The SCN master clock uses its connections to other brain regions to set all the other clocks in the body. One of these connections is to the pineal gland, which secretes the sleep hormone melatonin. Melatonin levels increase at night, and this is an important signal that helps synchronize the body's clocks by telling them when the sun has set. Connections from the SCN to other brain regions—including the arousal system and the VLPO sleep center—influence the circadian rhythms of sleep, physical activity, metabolism, and eating.[138] These connections explain why we tend to sleep at night and not during the day.

Since blue light controls melatonin secretion, melatonin levels at night

---

137 Hat tip to Deanna Arble for the orchestra metaphor.

138 The subparaventricular zone and the dorsomedial hypothalamus are particularly important for this.

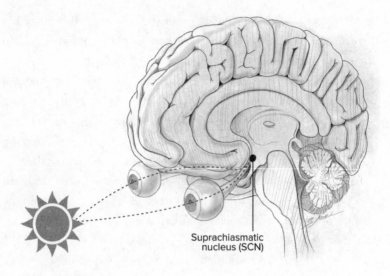

Figure 43. The suprachiasmatic nucleus and its input from the retina.

are exquisitely light sensitive. Artificial blue light, such as that emitted by the lightbulbs, computers, tablets, televisions, and cell phones in your house, suppresses melatonin levels at the time of night when it should be increasing. Your body's thirty-seven trillion cells don't get the message that it's night-time until you turn out the lights—several hours after the sun goes down. This pushes your biological wake-sleep cycle back by a few hours relative to the day-night cycle of the sun, desynchronizing the two. When it's time to go to sleep, your body isn't ready yet, and it also isn't ready when it's time to get up. This undermines your sleep quality and next-morning wakefulness. The system that evolved over billions of years to synchronize the body's daily rhythms to the cycle of the sun is easily fooled by modern technology.

Circadian disruption can do more than just interfere with our sleep, however. Research is increasingly suggesting that it's one of the lesser-known reasons why many of us grow fat and sick.

SYNCHRONIZING SLEEP

Technology such as lightbulbs undermines our sleep quality and perhaps our health, but technology can also provide solutions. There are two keys to keeping the circadian rhythm properly synchronized. The first is to reduce light intensity and blue-spectrum light at night. There are many ways to accomplish this. One easy fix is to get rid of night-lights and other sources of nighttime light in the bedroom, ensuring that it's completely dark while you sleep. Another easy way is to get rid of "full-spectrum," "daylight," and "cool white" lightbulbs in your house, replacing them with "warm white" bulbs that emit less blue light (or limit full-spectrum bulb use to daytime). Look for a color temperature of 3000 K or lower on the box. Dimmer switches can help by allowing you to match light intensity to the time of day. Turning down the brightness of your light-emitting devices, such as televisions, tablets, and smartphones, after sundown can also help. On your computer, you can download a neat utility called f.lux that automatically changes the color spectrum of your monitor to a warmer hue at sundown. Similar apps, such as Twilight, are available for smartphones and tablets. And finally, you can buy glasses that specifically block blue-spectrum light without blocking other wavelengths. These are available from a variety of online retailers, but one popular and inexpensive option is the blue-blocking SCT-Orange safety glasses manufactured by Uvex. Studies have shown that blue-blocking glasses completely eliminate the effects of artificial light on melatonin levels, suggesting that they help the SCN master clock understand that it's nighttime. Blue-blocking glasses let you use your lightbulbs and electronic devices as usual, while reducing their impact on your circadian rhythm. And you get to wear sunglasses at night.

The second key is exposing yourself to bright, blue-spectrum light early in the day so the SCN gets a clear message that it's

Continued on next page

Continued from previous page

daytime. In conjunction with limiting blue-spectrum light at night, this helps keep the circadian clock in the proper phase relative to the sun. The best way to get bright, blue-spectrum light is to do what our ancestors did: go outside. Even on a cloudy day, sunlight is much brighter than indoor light, and it tends to contain more blue-wavelength light than most lightbulbs. If you can't go outside, bright, full-spectrum indoor lighting is a good alternative.

CIRCADIAN CACOPHONY

Many researchers become interested in a subject because it affects them personally, and Deanna Arble, a postdoctoral researcher at the University of Michigan, is no exception. "When I was in high school," she recalls, "I could sleep twelve hours a day, every day, and still take naps. I was an outlier." As an undergraduate neuroscience major at the University of Virginia, she was drawn to sleep research because there was still so much that wasn't known about sleep. She approached an established circadian rhythm researcher named Michael Menaker and interviewed for a position in his lab. During the interview, Menaker noted that his research often involves working with mice in the dark. When Arble asked how they do research in the dark, he explained that they use night-vision goggles. Arble was sold. "I got into circadian research for the night-vision goggles," she jokes, "but I stayed for the science." Arble's subsequent Ph.D. work with Fred Turek at Northwestern University provided critical evidence that a disrupted circadian rhythm might contribute to our expanding waistlines.

To understand the importance of Arble's work, first we have to consider studies that have examined the health effects of rotating shift work in humans. In rotating shift work, people work at different times on different days, including at night, and they are typically unable to maintain a regular sleep schedule. Their bodies' circadian rhythm is perpetually misaligned with their cycle of activity and light exposure, and so they often sleep, eat, and perform physical tasks during times of day that their master clock isn't anticipating.

If circadian biology plays an important role in adiposity and health, then rotating shift workers should be heavier and less healthy than people who consistently work during the day. Many studies have shown that this is the case. Rotating shift work is associated with an alarming array of health problems, including obesity, type 2 diabetes, cancer, and cardiovascular disease. The longer a person works a rotating shift, the more likely they are to gain weight and become ill. Why?

This brings us back to Deanna Arble and her mentor Fred Turek. They were familiar with the growing body of research on rotating shift work, and they knew that digestive and metabolic functions follow a circadian pattern. Yet they also knew an additional critical detail: Not only can the master clock desynchronize from the day-night cycle of the sun, but individual organ clocks can desynchronize from one another. For example, the SCN master clock can become desynchronized from the clocks in the intestine, the liver, the pancreas, and other organs that regulate digestion and metabolism. That's because the latter group of clocks can be set by meal timing, and as far as they're concerned, the meal signal is stronger than the one coming from the SCN. Usually, this isn't a problem because the meal signal and the SCN signal are more or less aligned when we eat during the day. The conductor is in charge, and the orchestra is playing in harmonious synchrony. Yet when we disrupt the circadian rhythm by eating late at night, traveling to a faraway time zone, or doing shift work, some of our organ clocks can fall out of sync with others. What would normally be a harmonious performance is reduced to a disorganized cacophony. Researchers call this *circadian desynchrony* and hypothesize that it leads to metabolic problems and weight gain.

To test this idea, Arble studied two groups of mice, both eating the same fattening diet. One group had access to food only during the twelve hours of the day when they would normally be awake and active, while the other only had access to food during the twelve hours of the day when they would normally be asleep.[139] Under normal, unrestricted conditions, mice eat about two-thirds of their food during the wake cycle and one-third during the sleep cycle. In Arble's experiment, however, they could only eat during one period or the other, leading to two circadian conditions: one in which digestive and metabolic clocks were synchronized with the master clock in the brain, and one in which they were desynchronized.

Surprisingly, mice in both groups ate the same amount of food. Yet after two weeks, a striking trend began to emerge: the synchronized mice were hardly gaining weight, while the desynchronized mice were gaining fast. By the end of the six-week experiment, the desynchronized mice had gained nearly two and a half times more weight than the synchronized mice, and they tended to have more body fat as well. They were eating the same number of calories but metabolizing them differently.[140]

Subsequent rodent studies from other research groups have generally confirmed and extended Arble and Turek's findings. Together, they lead us to two surprising conclusions: The first is that circadian desynchrony accelerates weight gain when rodents are fed a fattening diet. The second, even more striking conclusion is that fattening rodent diets aren't particularly fattening when they're only available during the appropriate time of day! This suggests that what is really fattening in these experiments is not simply fattening food but the combination of fattening food *and* circadian desynchrony.

OK, this is interesting, but how relevant is it to humans, since most of us don't get up in the middle of the night to eat? Well, first of all, some of us *do* eat in the middle of the night. The most extreme example of this is a disorder called *night eating syndrome,* in which people eat the most of their calories at night, including in the middle of their sleep period. These people tend to be somewhat heavier than average. But we don't necessarily have to eat in the middle of the night to experience the negative effects of circadian disruption. Remember that most people living in affluent nations already have shifted, desynchronized, or otherwise disrupted circadian

---

139 Rodents have a reversed circadian cycle compared to humans; that is, they're active when it's dark and they sleep when it's light. The rodent SCN is sensitive to light just like ours, but its downstream effects on other clocks are reversed.

140 This result implies that calorie expenditure may have decreased in the desynchronized group.

rhythms due to artificial light at night, an insufficiently dark bedroom during sleep, a lack of morning sun, jet lag, and/or shift work. It's likely that we could stay slimmer and healthier with less effort if we took better care of our circadian clocks.

9

LIFE IN THE FAST LANE

J en slams her car door and rushes toward the office building where she works, nearly tripping over a crack in the sidewalk as she checks her watch. She spent forty-five minutes in a traffic jam, and now she's almost late for a meeting with her boss. She reviews her talking points to herself as she strides down the office hallway. Jen likes her boss—and her job—but today is her yearly performance evaluation, and she's nervous about it. She arrives just in time for the meeting, which goes about how she expected. The feedback is good overall, but she gets chastised for a careless mistake she made with a client recently. Her job is secure, for now. Although at times stress pushes Jen to perform her very best, at other times, it undermines her performance, saps her quality of life, and drives her to overeat unhealthy food.

Psychological stress is nearly ubiquitous in the modern world. Careers, finances, home ownership, parenthood, chronic health problems, social isolation, and many other life stressors add up to a potent brew for many people. Since 2007, the American Psychological Association has used nationwide surveys to track many aspects of stress and published the results in its yearly *Stress in America* reports. These reports show that three-quarters of Americans regularly suffer negative physical or psychological effects from stress, such as headaches, fatigue, upset stomach, insomnia, and irritability.

Are we maladapted to our current level of stress because our ancestors led more serene lives than we do today? Although it makes a good story to say that modern cultures suffer from more chronic stress than our distant ancestors did, convincing evidence is hard to come by. It seems to me that current nonindustrial cultures have plenty to worry about, including high infant mortality rates, deadly infectious diseases, accidents, homicide, starvation, witchcraft, and sometimes even predation.[141] Perhaps we've lost our traditional coping strategies, like close community bonds and daily physical activity. In any case, whether or not we're more stressed than our ancestors, stress is a powerful force in the modern world that makes a lot of people miserable.

Given the profound effects stress exerts on many brain systems, it's no surprise that it impacts eating behavior. What is surprising, however, is that the textbook response to stress is actually a *loss* of appetite, reduced food intake, and weight loss. This is certainly true for many types of stress, particularly physical stress—such as the flu—and powerful, acute psychological stress resulting from things like a car accident. But the full story is more complicated. This complexity is well illustrated by the 2007 *Stress in America* report, which includes data on how Americans react to and cope with stress in their lives. One of the most commonly reported effects of stress was a change in eating behavior—79 percent of people reported such a change. However, people reported conflicting effects. Forty-three percent reported overeating in response to stress, while 36 percent reported skipping a meal, and this result has been repeatedly confirmed by other studies. So it appears that stress can cause you to overeat or undereat, depending on who you are and what type of stress you experience.

In my view, this is an extraordinary finding that begs us to dig deeper. It would make sense if we all overate, to varying degrees, in response to stress. But to see diametrically opposed responses in different people is just odd.

---

141 In general, the per capita risk of all these things I listed is much higher among nonindustrialized societies than in modern affluent nations such as the United States. Witchcraft and other supernatural threats may not be rational fears, but they nevertheless cause much anxiety in traditional cultures.

Solving this puzzle will require us to push into the neuroscience and psychology of stress.

FEARLESS MONKEYS

What is stress? Broadly, it's a coordinated set of physiological and behavioral responses that the brain engages to meet a challenging situation.[142] These responses limit the damaging consequences of a threatening scenario and make us perform our best when the stakes are high. For many animals, stress is a way of life. From avoiding predators, to finding scarce food, to competing for mates, the ability to mount a fast and effective stress response is critical to survival and reproduction in a natural environment. Because of this, stress responses are deeply wired into many of our brain systems.

Yet stress isn't just one thing. For example, if you get into a car accident and you're losing blood quickly, that's a different kind of stress from when you're in a meeting with your boss that could end with a pink slip. The first scenario is an example of stress arising from physical damage to your body, while in the second, the stressor is an abstract concept. You're stressed about the uncertain prospect of losing your job, and all the downstream effects that might have on your future life, such as defaulting on your mortgage or rent, having to explain the situation to your partner, and having to go through the job search rigmarole. This is *psychological* stress, and it's the type that affects us the most in affluent nations today. Certain types of psychological stress have the peculiar ability to provoke overeating.

Ultimately, although different types of stress provoke different brain responses, they all engage a partially overlapping collection of brain circuits I'll call the *threat response system*. The threat response system activates a number of processes that together serve to protect you from danger and help you make the most of a challenge.

In 1939, an unusual experiment by the German American psychologist Heinrich Klüver and the American neurosurgeon Paul Bucy took the first steps toward unraveling the core brain circuitry of the threat response

---

142 There are also forms of stress that aren't managed by the brain, including several types of cellular stress, but they are beyond the scope of this book.

system. As part of an investigation into the effects of the psychedelic drug mescaline, Klüver obtained a group of rhesus monkeys and arranged for Bucy to surgically remove specific parts of the monkeys' brains, including the temporal lobe of the cortex, the hippocampus, and an area called the amygdala. He called these surgically transformed animals "bilateral temporal monkeys." Although the surgery didn't have much impact on the monkeys' response to mescaline, it produced pronounced behavioral changes. Most relevant to us is the fact that it made them fearless. According to Klüver and Bucy:

> The typical reaction of a "wild" monkey when suddenly turned loose in a room consists in getting away from the experimenter as rapidly as possible. It will try to find a secure place near the ceiling or hide in an inaccessible corner where it cannot be seen. If seen, it will either crouch and, without uttering a sound, remain in a state of almost complete immobility or suddenly dash away to an apparently safer place. This behavior is frequently accompanied by other signs of strong emotional excitement. In general, all such reactions are absent in the bilateral temporal monkey. Instead of trying to escape, it will contact and examine one object after another or other parts of the objects, including the experimenter, stranger or other animals . . . Expressions of emotions, such a vocal behavior, "chattering" and different facial expressions, are generally lost for several months. In some cases, the loss of fear and anger is complete.

These monkeys also failed to mount a normal fear reaction in threatening social situations, leading them to fight with larger, more dominant monkeys and often sustain severe injuries. This odd collection of behaviors was dubbed *Klüver-Bucy syndrome*. Later studies found that the fearlessness of Klüver-Bucy syndrome could be largely replicated simply by disrupting the amygdala and adjacent brain tissue.

In the seventy-five years since Klüver and Bucy's seminal findings, researchers have gradually uncovered a complex network of brain structures that play a role in the threat response system. These structures range from the oldest parts of the brain stem to the most recently evolved parts of the frontal cortex, emphasizing how deeply rooted the threat response system is in the brain, and its critical importance to the survival of our ancestors.

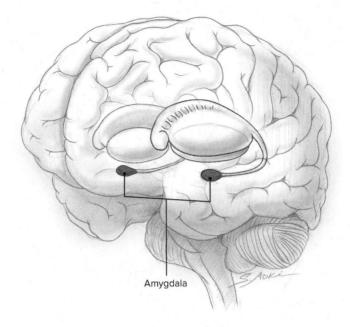

Figure 44. The amygdala.

At the heart of this network of brain structures lies the amygdala—roughly the size and shape of a large almond (see figure 44).[143] (The amygdala and other closely associated brain areas are often lumped together under the term "extended amygdala." Unless otherwise stated, when I say "amygdala," I'm referring to the entire extended amygdala region.)

Decades of research in animal models and humans suggest that the amygdala is the central node of the threat response system that responds to external and psychological threats, including the everyday psychological stressors we commonly call "stress." In broad strokes, here's how it works:

The amygdala cooperates with many different brain regions, both conscious and nonconscious, to scan for signs of a threat. Some of these regions process concrete sensory information, such as objects moving rapidly toward you, things that look like spiders, or loud sounds, while others process abstract concepts like being laid off, carrying debt, or arguing with a

---

143 It gets its name from the ancient Greek word for *almond*.

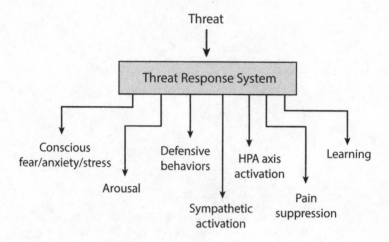

Figure 45. The functions of the threat response system.

loved one. When the amygdala detects a threat, it communicates with other brain regions to activate a coordinated suite of threat responses designed to minimize the potential damaging consequences of the situation. This is illustrated schematically in figure 45.

The amygdala increases your level of arousal by activating many of the same brain regions that are involved in regulating the sleep/wake cycle (see figure 41). This bathes broad swaths of your brain in dopamine, serotonin, noradrenaline, and other chemicals that focus your mind on the problem and motivate you to do what it takes to resolve it. That's why it's hard to sleep when you're stressed.

Depending on the threat, the amygdala may send signals to your brain stem that activate fast defensive reflexes, such as startling, freezing, protecting your head, or closing your eyes. These signals can also instinctively generate facial expressions of fear or anxiety—recognizable across all human cultures.

At the same time, the amygdala sends signals to activate your sympathetic nervous system. The sympathetic nervous system is a network of nerve fibers that runs throughout the body and participates in the fight-or-flight response (see figure 46). Your pulse and breathing rate quicken, and your blood pressure rises. Your palms begin to sweat. Digestion slows, and in extreme cases, your bladder and rectum may expel their contents. Blood

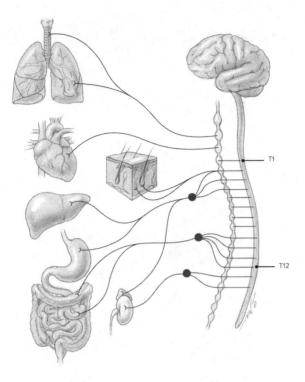

Figure 46. The sympathetic nervous system. This simplified illustration focuses on the threat response pathways of the sympathetic nervous system.

flow to your muscles increases. The levels of sugar and fat in your bloodstream begin to climb, providing your muscles with more energy for fight or flight. This process prepares your body for vigorous action, but it happens even in response to psychological threats that don't require you to meet physical challenges. Your body is preparing to fight off a grizzly bear, even if the actual threat is an Excel spreadsheet.

Simultaneously, the amygdala activates a critical part of the threat response system called the *hypothalamic-pituitary-adrenal axis* (HPA axis). It does so by sending a signal that causes the hypothalamus to release a chemical called corticotropin-releasing factor (CRF).[144] CRF and a few related molecules turn out to be key players in the threat response.[145]

---

144 Specifically, the paraventricular hypothalamic nucleus.

145 These related molecules include the urocortins.

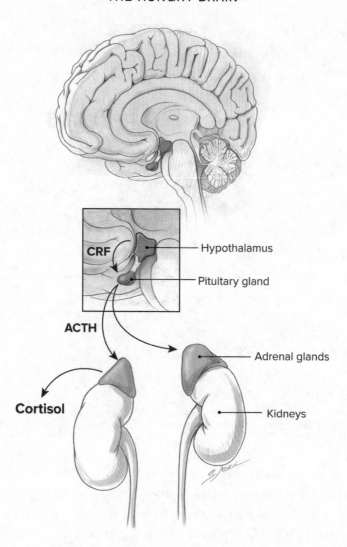

Figure 47. The hypothalamic-pituitary-adrenal axis. The hypothalamus produces corticotropin-releasing factor (CRF), which stimulates the pituitary gland to secrete adrenocorticotropic hormone (ACTH), which stimulates the adrenal glands to secrete the stress hormone cortisol.

The release of CRF is the first step in the signaling cascade of the HPA axis, which culminates in your adrenal glands producing the stress hormone cortisol (see figure 47). Some of the effects of cortisol, such as increasing blood levels of sugar and fat, are similar to those of the sympathetic nervous system, but on a slower, longer time scale. These sustain the high metabolic demands of dealing with a stressor for a long time. Other effects of cortisol include suppressing immune function and, as we will soon see, increasing food intake. Because of its slow kinetics, cortisol is a key player in the long-term response to chronic stress.

The amygdala also makes its own CRF, which plays an important role in increasing your arousal level, your anxiety-related behaviors, and the activity of your sympathetic nervous system. CRF stimulates the brain as a whole to shift from normal, everyday behaviors like eating and socializing, to threat response behaviors like running away and (perhaps) thinking about how you'll pay the bills. To bring it back to the basal ganglia, CRF seems to increase the bid strength of threat-related option generators, and weaken the bid strength of option generators that are unrelated to the threat. CRF is such a powerful mediator of stress that several pharmaceutical companies are investing large amounts of money into trying to block it in humans.[146]

The final act of the brain's threat response is learning. Once you've survived a threatening situation, your amygdala learns how to recognize and respond to similar situations more effectively in the future. With practice, your brain gets better at protecting you. But even though the threat response system evolved to keep us safe—and it often does—at times it can do things that seem quite counterproductive, like making us eat too much.

THE EVERYDAY STRUGGLE . . .
OF A RHESUS MONKEY

At the Yerkes National Primate Research Center in Atlanta, Georgia, neuroscience researcher Mark Wilson observes five dun-colored female rhesus monkeys in an enclosure. On one side of the enclosure, two monkeys are seated peacefully, one grooming the other. On the other side, one monkey suddenly slaps another on the side of the head and, a moment later, chases the slapee into a corner, where it cowers until the aggressor relents. The fifth monkey watches the spectacle and then saunters over to a feeding station, where it reaches into a tube in the center of a complicated-looking

---

146 Some compounds already exist, such as antalarmin, which blocks the CRF-1 receptor.

machine, pulls out a food pellet, and eats it. Unbeknownst to the monkey, there is a tiny electronic tag implanted in its wrist that the feeding station reads each time it reaches into the feeding tube for a pellet. The machine also precisely weighs each pellet, and in this way it accurately registers each monkey's food intake. These data are helping Wilson and his team understand how, and when, stress makes us overeat.

How can rhesus monkeys teach us about stress eating? Like many animals, rhesus monkeys spontaneously organize into social hierarchies, and dominant monkeys maintain the hierarchy by harassing and occasionally hitting or biting subordinates. That means Wilson's team doesn't have to do anything to create stress: The monkeys do it to themselves. "The primary stressor," explains Wilson, "is just getting harassed, day in and day out." Most of the harassment in stable social hierarchies is what Wilson calls *noncontact aggression*—the monkeys are threatening and chasing one another but not making physical contact. This isn't too far off from the chronic psychological stress most humans experience, where the stressor is most often a *threat* of harm (e.g., a fear of being laid off) rather than harm itself (e.g., actually being laid off). Even though no one gets hurt, it's still unpleasant. And it's uncontrollable.

This lack of control is a key factor in modeling the most harmful types of human psychological stress. In the modern world, we're often subjected to stressful events we can't easily control, like traffic, bullying, nagging, illness, deadlines, and debt.[147] Research in psychology and neuroscience suggests that uncontrollable stressors have a stronger effect on the threat response system, and are much more harmful to our health and mental state, than stressors we believe we can control.

How do these uncontrollable, often social, stressors affect food intake? Remarkably, it depends entirely on the food that's available. When Wilson's team feeds its monkeys healthy, unrefined, high-fiber chow, stressed subordinate monkeys eat less and lose weight while dominant monkeys maintain weight. Yet when the researchers give the monkeys a choice between standard chow and a very rewarding high-fat, high-sugar diet, the

---

147 We could debate whether or not these stressors are genuinely uncontrollable, but what actually matters to the brain is whether or not they're *perceived* as uncontrollable. What seems controllable to one person may seem uncontrollable to another.

monkeys' eating behavior changes dramatically. First of all, not surprisingly, both dominant and submissive animals prefer the rewarding diet and eat it at the expense of the healthy diet. Yet the dominant animals keep eating the same amount of food as before. In contrast, the stressed subordinates *double* their daily calorie intake. So in the context of a strict healthy diet, stress makes monkeys undereat, whereas when they have a choice between healthy fare and junk food, they overeat spectacularly.

Following up on this finding, when Wilson's team blocks the effects of CRF in the monkeys' brains, knocking out a key component of the threat response system, the subordinate monkeys stop overeating. This confirms that overeating in stressed monkeys, and perhaps ourselves as well, is a direct result of the activation of the threat response system in the brain.[148]

Wilson's research suggests that the magic formula for overeating is the combination of chronic uncontrollable stress and a choice of highly rewarding food.[149] This specificity may go a long way toward explaining why some people overeat when they're stressed and others don't. Each person experiences a different combination of stressor type and food environment, and only some of these combinations are the magic formula.

This may also explain part of the reason why traditionally living cultures don't seem to overeat and gain weight when they're chronically stressed, like many of us do. Even though traditional cultures have plenty to be stressed about, their food tends to be much simpler, less refined, and less rewarding. They may be more like the stressed monkeys who don't overeat because they only have access to healthy, unrefined food. In contrast, many of us are like the stressed monkeys who overeat because we live in an extravagant food environment.

Why does chronic uncontrollable stress push us to overeat, and why does this only seem to happen when there's highly rewarding food around? To answer these questions, we'll have to turn to the effects of uncontrollable stress on the endocrine system.

---

148 It also paints a complex picture, because CRF is classically thought to reduce food intake. It seems we still have a lot to learn about how the threat response system works.

149 Particularly when the monkeys have a choice between healthy and unhealthy diets.

HORMONE HUNGER

In 1910, the neurosurgeon Harvey Cushing saw a twenty-three-year-old patient named Minnie G. who was suffering from a peculiar and distressing disorder. Minnie G. had stopped menstruating, showed excessive hair growth, and most important for our purposes, she had abdominal obesity. Because of the fact that she also had increased fluid pressure in her brain case (hydrocephalus), Cushing suspected that her disorder was caused by a tumor in her pituitary gland that had disrupted her hormone levels. After seeing more patients with a similar ailment, Cushing determined that pituitary tumors were indeed often the cause. These tumors increase the volume of the portion of the gland that secretes adrenocorticotropic hormone (ACTH), a critical part of the HPA axis that mediates part of the body's stress response (see figure 47). In turn, excess ACTH tells the adrenal glands to secrete too much of the stress hormone cortisol, which causes the disorder that was eventually named Cushing's disease. More generally, it's called *Cushing's syndrome* when a person shows signs and symptoms of excess cortisol but we don't know the specific cause. Cushing was unable to perform an autopsy on the brain of Minnie G., so we'll never know whether or not she actually had a pituitary tumor, but today we do know that most people who spontaneously develop high cortisol levels do so precisely because of such tumors.

Yet there's another cause of Cushing's syndrome, which helps us shed more light on the connection between cortisol and overeating: This form is actually caused by medical treatment. Doctors often prescribe drugs similar to cortisol (such as prednisone) due to their remarkable ability to suppress the immune system. This can be useful if a person suffers from severe immune-mediated problems, such as rheumatoid arthritis or organ transplant rejection. Yet when taken at high doses, these drugs cause people to develop the abdominal obesity characteristic of Cushing's syndrome.

In 1987, Eric Ravussin and his research team set out to learn why this happens. They recruited twenty healthy young men, assigned them to either methylprednisolone (a cortisol-like drug) pills or placebo, and carefully monitored each group's food intake. Over a four-day period, Ravussin's team found that the methylprednisolone group ate a whopping 1,687

Calories more per day than the placebo group. This showed unequivocally that engaging a key part of the threat response system, albeit to an abnormally strong degree, causes overeating. Ravussin's team concluded:

> Our data suggest that therapeutic doses of [cortisol-like drugs] induce obesity mostly by increasing energy intake, an effect which may be related to the ability of [cortisol-like drugs] to act directly or indirectly on the [brain] regulation of appetite.

To understand why cortisol can have such a profound effect on food intake and adiposity, we need to return to the system in the brain that regulates adiposity and appetite, the lipostat. As we discussed previously, leptin tells the hypothalamus how much fat the body carries, and the hypothalamus uses that signal to regulate food intake and energy expenditure. When the hypothalamus becomes resistant to the leptin signal, adiposity increases.

With this in mind, there's a tidy explanation for why overeating and obesity are key features of Cushing's syndrome: Cortisol and related compounds cause leptin resistance in the hypothalamus. In 1997, a Swiss research team led by Katerina Zakrzewska determined that removing cortisol from the circulation of rats makes them exquisitely leptin-sensitive, and also lean (technically, these studies manipulated levels of corticosterone, the rodent equivalent of cortisol). Incrementally increasing their cortisol levels makes the rats incrementally less leptin sensitive and incrementally fatter. Research from other groups showed that cortisol-like compounds interfere with the ability of leptin to activate its signaling pathways in cells of the hypothalamus, and increase levels of the hunger-promoting substance NPY.[150] Together, this research shows that a key component of the threat response system can interfere with a key component of the lipostat, driving the nonconscious parts of the brain that regulate adiposity and appetite to favor overeating and fat gain.

Although this is interesting stuff, we haven't quite made it from saying that cortisol *can* cause leptin resistance, overeating, and abdominal

---

150 Specifically, cortisol-like compounds (glucocorticoids) inhibit STAT3 phosphorylation by the leptin receptor, a critical signaling event that mediates much of leptin's impact on energy balance in the hypothalamus.

fat accumulation in extreme cases, to saying that it *does* cause these things in people living their everyday stress-filled lives. The latter is much harder to demonstrate, and there is no smoking gun yet. However, there is quite a bit of evidence that's strongly suggestive.

Several large studies have shown that people with higher stress levels tend to gain more body fat over time than people with lower stress levels—and this fat gain occurs particularly in the abdominal area, resembling a mild form of Cushing's syndrome (they also exhibit metabolic changes that are reminiscent of Cushing's). Elissa Epel, professor of psychiatry at the University of California, San Francisco, believes that this phenomenon may be more prevalent than we think. "It's really insidious and common, and we often don't see it," explains Epel, because people can accumulate abdominal fat while still appearing fairly lean overall. These people are at a high risk of health problems because fat inside the abdominal cavity (where your digestive organs are) is more dangerous than fat under the skin, but they may be hard to identify because they aren't always classified as obese on the body mass index scale that doctors commonly use to diagnose obesity. At any level of overall adiposity, chronic stress seems to shift the distribution of body fat toward the visceral cavity, where it tends to wreak metabolic havoc.

Epel's group has found that some people secrete high levels of cortisol in response to experimental stressors under lab conditions, while others don't secrete much cortisol at all. The high secretors tend to eat more food when they're stressed, while the low secretors don't. Together, this paints a fairly coherent picture suggesting that cortisol may be a key reason why everyday stress can cause overeating and fat accumulation around the midsection. And it also helps explain why only certain people overeat in response to stress.

As it turns out, uncontrollable stress—like being hassled all day by four rhesus monkeys or your boss—has a particularly potent cortisol-raising effect. In contrast, when you're facing a challenge but you have a chance to determine your own fate, the situation feels less threatening, and your cortisol response is proportionally smaller. This may be part of the reason why uncontrollable stress is the most effective at driving overeating and adiposity.

Let's return to Jen's story, which we began at the opening of the chapter. Before and during her workday, Jen's amygdalae were powerfully activated by brain regions that understand the abstract concepts of traffic jams,

being late for work, and performance evaluations—and all their potential consequences. As her amygdalae stimulated her sympathetic nervous
system and HPA axis, her heart rate accelerated, her blood levels of sugar
and fat increased, her palms began to sweat, and she became more alert.
Jen doesn't know it, but she tends to secrete a lot of cortisol when she's
stressed—particularly when the stressor feels outside of her control. On
her way to work, she faced a maddening traffic jam, and there was nothing she could do but wait it out. When she walked into her boss's office,
she didn't know what he was going to say, but again there wasn't much
she could do to change the outcome. In the end, she made it to work on time,
and the evaluation went fine, but by that time her cortisol levels were soaring. This cortisol traveled to her brain, causing her hypothalamus to become less sensitive to the appetite-restraining effects of leptin. At her meals
that day, she noticed that her appetite was unusually large. Each time
this happens, Jen gains a bit of weight, especially around her midsection.

Yet there's another, more common-sense reason why some of us overeat when we're stressed, and research is increasingly suggesting that it
could be important: Junk food simply makes us feel better emotionally.

THE COMFORT FOOD CONNECTION

"When I first got into HPA axis research," reminisces Mary Dallman, professor emeritus of physiology at the University of California, San Francisco, "people didn't know what the hell was going on." Dallman waded
into physiology in the 1960s—a time when our understanding of HPA axis
biology was superficial, and the field was virtually off-limits to women.
She began as a technician at Harvard and eventually became the first
female research faculty member at UCSF. Her subsequent work played a
critical role in uncovering how the HPA axis is regulated—in particular,
how it shuts itself off.[151]

Yet in recent years, Dallman's research has expanded from studying
basic HPA axis biology to include matters of more direct relevance to
human health. As is often the case in the research world, Dallman explains

---

151 Another example of negative feedback. Cortisol acts in the brain to reduce HPA axis activation.

that "almost everything I've gone after has been from an accidental observation." One of these came from her husband, Peter Dallman, who happened to be a pediatric hematologist, a doctor who treats blood disorders in children. He frequently used dexamethasone, a cortisol-like drug, to treat specific blood disorders that benefited from suppressing the immune system. "He said the first thing you see when you give dexamethasone," recalls Mary Dallman, "is that the kids would start eating like crazy—and you knew the treatment was going to work." This led her to study the effect of cortisol and related compounds on calorie intake in rodents.

Although this was a productive line of research, Dallman knew there must be more to the story. Human studies were showing that stress doesn't just change the *amount* of food we eat; it also markedly alters the *types* of food we eat. Depending on how you're wired, calorie intake can go up or down during stress, yet regardless of this, most of us gravitate toward calorie-dense "comfort foods" like chocolate, ice cream, macaroni and cheese, potato chips, and pizza. Dallman suspected that people might actually be using food to self-medicate their stress, dampening the activity of their own threat response systems.

To test this hypothesis, Dallman's team turned to a tried-and-true model of stress in rodents called *restraint stress*. In this model, researchers place rodents in a small confined space for a short period of time, which activates their threat response systems. Researchers can then measure this activation by looking for increased activity of the sympathetic nervous system and the HPA axis, as well as defensive behaviors such as freezing and hiding.

To study whether tasty foods can dampen the stress response, Dallman's team gave one group of rats unlimited access to sugar water for ten days, while another group only got plain water. At the end of this period, researchers subjected both groups to restraint stress and then measured the stress-induced activation of their HPA axis. As predicted by Dallman's comfort food hypothesis, the group that drank sugar water showed a smaller stress response than the group that drank plain water. It appeared that sugar helped them feel better in the face of stress.[152] Yet it's not just sugar: Dallman and others have since shown that giving rats access to a high-fat food does the same thing.

---

152 I'm using the phrase "feel better" loosely here. It's difficult to know how a rat actually feels.

What is it about comfort food, exactly, that dampens the stress response? Is it the body's metabolic response to sugar, fat, and/or calories, or is it the food's reward value itself? Yvonne Ulrich-Lai, a neuroscience researcher at the University of Cincinnati, set out to answer this question. She and her team offered rats intermittent access to small volumes of sugar water, water sweetened with the calorie-free sweetener saccharin, or plain water, and then measured the activation of their HPA axis following restraint stress. She hypothesized that sugar's stress-busting effect is due to its metabolic impact on the body, not its rewarding effects in the brain, and therefore that saccharin would be ineffective. "The result," explains Ulrich-Lai, "was completely opposite my hypothesis. The saccharin worked just as well as sugar." Subsequent experiments confirmed that the *sweet taste itself* was responsible.

In 2010, her team went further by seeing if rewarding behaviors besides eating also dampen stress responses. To do this, they deployed the only natural reward that can compete with tasty food: sex. Ulrich-Lai's team gave one group of male rats daily access to a sexually receptive female and let another group of males see and smell a female but not touch her. Then they tested the rats' stress responses to restraint stress. The sex worked. In fact, according to Ulrich-Lai, it was a little bit more effective than the sugar.

Peering into the rats' brains, her team was also able to determine that natural rewards may attenuate stress by changing how the amygdala processes stress-related information. Together, Ulrich-Lai's results suggest that rewarding behaviors directly oppose the activation of the threat response system in the brain. Although the finding still needs to be confirmed in humans, it offers a compelling explanation for why we seek highly rewarding junk food (and drugs like alcohol) when we're stressed: The reward value itself helps us feel better by dampening the activity of the threat response system. It may also explain why stress only causes overeating when there's highly rewarding food around: When we want to self-medicate our stress, bland food just doesn't cut it.

Let's return to Jen's story once again. Like the rest of us, Jen doesn't like to feel stressed, and she takes steps to manage it. She knows that when she's stressed, eating something tasty helps her feel better. On her way home from work, she stops by the grocery store to buy food for the next few days. In addition to the (mostly healthy) food she puts into her cart, she grabs a

CONSTRUCTIVE COMFORT

Ulrich-Lai's findings have a valuable practical implication: The brain contains a natural stress relief pathway that we can tap into using a variety of everyday behaviors. "If we can understand that [built-in] stress relief pathway," explains Ulrich-Lai, "then we might be able to find other ways to target it, either behaviorally or using drugs." Why would we want to find other ways to target it besides comfort food? Because eating calorie-dense comfort food too often can make us gain weight and degrade our health. If we might gain the same stress-busting benefit from any rewarding behavior, then why not choose something more constructive, like calling a friend, jogging, gardening—or romance?

box of chocolate chip cookies. Between her heightened hunger and her desire for comfort food, she ends up eating a third of the box—far more calories than she needs. She resolves that the next time she feels that way, she'll take a walk or a bubble bath instead.

10

THE HUMAN COMPUTER

The human brain is a three-pound pink gelatinous mass of tissue. It bears little resemblance to the hard metal and plastic components inside your computer. Yet they both perform the same basic function: information processing. Although they differ in many details of operation, brains and computers both collect information inputs, process them, and use them to generate useful outputs. I think this is a good starting point for understanding how the brain works.

The inputs to the brain come from two places: inside and outside the body. From outside, the brain receives information from our external sense organs, such as sight, hearing, smell, taste, and touch.[153] From inside, it receives information from many different sensors that convey information about the position of our limbs, how our head is accelerating and rotating, our core temperature, the quantity and quality of our gut contents, bladder and rectum fullness, our blood concentration of ions and glucose, the amount of fat we carry, digestive distress, infections, tissue damage, and countless other variables.

As in a computer, the brain processes this information and uses it to

---

153 The sense of touch is actually quite complex because it can detect many different characteristics of the environment, including pressure, vibration, cold, and heat.

generate useful outputs. Once again, these outputs affect two things: what happens inside the body and what happens outside the body. We call the first one physiology and the second one behavior. So your brain collects information from inside and outside your body and uses it to appropriately regulate what's happening inside and around you, in order to benefit you. For example, if you see a basketball rapidly approaching your head, you'll duck out of the way if you can. In that case, your brain collected information from your visual system and used it to generate movements that got you out of the way of the ball.

Here's a less intuitive example that relates more closely to our subject of interest. When you lose weight, your brain detects falling levels of leptin and increases your motivation to eat. In that case, your brain is using an internal input to regulate a behavioral output, and it happens largely outside of your conscious awareness. Here's the point I'm driving toward: The outputs of your brain, including your appetite and eating behaviors, are determined by the input cues it receives.[154] Some of these cues are processed by conscious circuits, and we're very much aware of how they affect us. Many others are processed by nonconscious circuits, which influence our physiology and behavior in ways that we have little awareness of, and little direct control over. Yet they can nevertheless have a substantial impact on our lives.

I've argued that these nonconscious circuits explain why we overeat, despite our best intentions. As Daniel Kahneman relates in his book *Thinking, Fast and Slow*, the brain's thought processes can be roughly divided into two systems. System 1 is fast, effortless, intuitive, and nonconscious, while system 2 is slow, effortful, rational, and conscious. System 2 understands the long-term consequences of our choices and represents our rational intentions: it wants you to eat the right amount of nutritious food, get plenty of exercise and sleep, and stay lean, fit, and healthy to a ripe old age.

Unfortunately, system 1 isn't so rational. It has its own agendas, crafted by millions of years of natural selection. It guides our behavior and physiology using a collection of hardwired heuristics that helped us survive and thrive in the world of our distant ancestors. It didn't evolve to guide us

---

154 Plus, of course, the current state of your brain (as determined by genetics, previous experience, and random chance).

through a world of credit cards, pornography, and addictive drugs. And it didn't evolve to handle easy access to calorie-dense, highly palatable food. It often disagrees with system 2, and it can be extremely persuasive. System 1 is why we overeat, despite the fact that we know it makes us fat and undermines our health and well-being.

Although Kahneman doesn't discuss in his book the brain structures that underlie systems 1 and 2, we know that system 1 represents more than one circuit. In fact, it's a hodgepodge of many circuits that perform a variety of processing tasks. In this book, I've taken you on a tour of some of these processes that drive us to overeat in the modern world, despite our best intentions.

First, we encountered the reward system, centered in the basal ganglia, which teaches us how to get food properties—such as fat, sugar, starch, and salt—that the brain instinctively views as valuable. The reward system collects food-related cues from our external sense organs and our digestive tract, and guides us toward valuable foods by helping us learn and motivating our behavior. What we experience is that we crave and enjoy certain foods more than others, and we develop deeply ingrained eating habits around the foods we enjoy. Unfortunately, this system evolved in an ancient world where calories were hard to come by, and obtaining them required a lot of work, and therefore motivation. In the modern affluent world, where calorie-dense, highly rewarding foods are ubiquitous, our hardwired motivation to eat remains strong, and it drives us to overconsume.

Next, we explored the economic choice system, centered in the orbitofrontal cortex and the ventromedial prefrontal cortex, which integrates the costs and benefits of possible actions and selects the one that's the best "deal." This system contains both conscious and nonconscious elements, because it integrates costs and benefits from all parts of the brain that relate to the decision at hand. Some of the costs and benefits it considers relate to system 2 processes, such as predicting the future impact on your waistline of eating a pastry, or calculating whether you can afford to buy it. Yet much of what it considers is nonconscious, and in many animals, including humans, it appears to be wired to value calories above all other food properties. And the easier the calories are to get, the more of them we eat. Another way of saying this is that when it comes to food, its primary cues are calories and convenience. This becomes a liability in a world where calorie-dense foods are more convenient than ever to purchase, prepare, and consume.

The lipostat is a third system, located primarily in the hypothalamus, which nonconsciously regulates adiposity by influencing appetite, our responsiveness to seductive food cues, and our metabolic rate. It takes its cues primarily from the hormone leptin, which is produced by fat tissue, although it also responds to food reward, protein intake, physical activity, stress, and possibly sleep (as well as other factors that are beyond the scope of this book). The lipostat has one job: to prevent your adiposity from decreasing. And it's very good at what it does, because in the world of our ancestors, losing weight meant having fewer offspring. This is the system that makes weight loss difficult and often temporary, and it may also be part of the reason why our weight tends to creep up over the years. It partially explains why people with obesity tend to eat more than lean people, making obesity a self-sustaining state. When the lipostat squares off with our best intentions to eat the right amount and stay slim, it usually wins in the end.

Working in parallel with the lipostat, the satiety system regulates food intake on a meal-to-meal basis by making us feel full and reducing our drive to continue eating after we've had enough. Located primarily in the brain stem, the satiety system takes its cues from the digestive tract, which relays how much volume we've eaten, and the protein and fiber content of our food. The satiety system also receives cues from the reward system, which tends to shut down the feeling of satiety when we eat highly rewarding foods, such as pizza, french fries, and ice cream. And finally, it takes important cues from the lipostat, which increases or decreases satiety to help maintain the stability of body fat stores. One of the reasons why modern food tends to be so exceptionally fattening is that it doesn't provide the cues the satiety system needs to appropriately regulate calorie intake. We tend to use the sensation of fullness as an intuitive signal that we've eaten enough, so when we eat calorie-dense, low-fiber, low-protein, highly palatable foods that provide little satiety per calorie, we overeat substantially without even realizing it.

The genetic details of how each person's lipostat and satiety system are constructed go a long way toward explaining why some of us develop obesity in the modern world, while others remain lean. Each person's brain has a unique genetic blueprint, and this nudges us to interact differently with food and defend different levels of adiposity. Some people are naturally resistant to overeating even in a very fattening food environment, while most of us are susceptible. Certain lucky folks don't gain weight even if they do overeat, because their lipostat just burns off the excess calories. This makes it very difficult to judge people for their weight, since we're all born with different predispositions. At the same time, for most of us, our genetic makeup is just a predisposition—not an inevitable fate. Our ancestors four generations ago carried the same genes we do, yet they rarely developed obesity because they lived in a different context that provided profoundly different cues to the brain and body.

The sleep and circadian rhythm systems, located in the hypothalamus, the brain stem, and other brain regions, nonconsciously influence eating behavior and adiposity in large part by interacting with the reward system, the lipostat, and the economic choice system. The amount of restorative sleep we get, the timing of our sleep, the blue-wavelength light we see, and the timing of our meals are key cues that regulate these systems. When we don't sleep enough, or don't sleep well enough, it increases the reward system's responsiveness to food cues, and we often end up eating more calories. It affects our economic choice system, promoting an "optimism bias" that makes us think more about the current benefits of eating a pastry than the future costs. And it nudges us to become less faithful to our rational, constructive daily goals, such as eating healthy food. When our sleeping and eating behaviors are misaligned with the day-night cycle of the sun, or with our own internal clocks, it can also cause us to gain weight and undermine our metabolic health. In the modern affluent world, with its time-consuming responsibilities, high rates of sleep disorders, abundant artificial light at night, stimulating media, rotating shift work, and travel between time zones, the sleep and circadian rhythm systems receive cues that often send our eating behavior and weight in the wrong direction.

Finally, the threat response system, rooted in many parts of the brain but coordinated in large part by the amygdala, is a largely nonconscious collection of processes that help us manage challenging situations by altering our behavior and physiology. This system takes its cues from a variety of sensory inputs that convey information about potential threats, such as vision and hearing, but also from abstract concepts, such as the possibility of being laid off. In the modern world, most of the threats it deals with are psychological, but in many ways it still reacts as if we're fighting off wild beasts. It's not clear that we have more to be stressed about today than our ancestors did, yet it's possible that we cope with our stress less effectively than we did in times past. In certain people, psychological stress sharply increases cortisol levels, and this may reduce the sensitivity of the lipostat to leptin, in turn increasing food intake and the accumulation of body fat. This is particularly true when we feel like we have little control over a stressful situation. Stress also shifts our eating preferences toward comfort food—which is usually calorie dense and highly palatable—because it helps dampen the activity of the stress response system and makes us feel better.

The brain is a complex organ, and it contains many circuits that influence eating behavior besides those I described in this book—and I'm certain that more remain to be discovered. Yet those we've explored in these chapters play key roles in determining our calorie intake and adiposity. These circuits respond to the cues they receive, and the modern affluent food environment provides a perfect storm of cues that promote overeating. I believe these circuits caused Yutala, the fattest man on the island of Kitava, to develop a body shape unfamiliar to his traditionally living community, after adopting a modern lifestyle in the small Papua New Guinea city of Alotau. I also believe these are the circuits that pushed our own culture from leanness to obesity, as our way of life departed radically from that of our ancestors.

TAKING THE NEXT STEPS

Where will the research go from here? As I hope I've conveyed, neuroscience is a vibrant field that is rapidly expanding our understanding of the human brain. Yet as the most complex object in the known universe, the brain has many secrets left to divulge. Within the sphere of eating behavior and obesity research, many scientists are continuing to explore the nonconscious circuits that drive us to overeat and gain fat. Much of this work revolves around identifying the input cues these circuits respond to, understanding the molecular details of the circuits themselves, and determining how each of these can be manipulated to prevent and reverse obesity.

One excellent example is the scientific quest to understand why certain types of weight-loss surgery are so effective at treating obesity. While many people with obesity struggle to lose fat using diet and lifestyle approaches, usually without much success, those who undergo specific surgical procedures, such as Roux-en-Y gastric bypass or sleeve gastrectomy, experience substantial and durable fat loss with much less effort.

Initially, surgeons assumed the obvious: These procedures work because they reduce stomach volume and digestive efficiency, so the recipient simply can't fit much food into his stomach and also ends up flushing calories down the toilet instead of absorbing them. But when researchers took a closer look, the story turned out to be much more interesting. People who undergo these procedures still have the digestive capacity to eat enough calories to sustain obesity, and they absorb calories nearly as well as they did before surgery. Yet they simply lose their desire to eat rich foods. While before surgery, they may have craved large portions of hamburgers, fries, and soda, after surgery they gravitate toward smaller portions of lighter fare, such as vegetables and fruit. And as they lose weight, they never show signs of the starvation response that usually accompanies substantial weight loss, implying that weight-loss surgery turns down the set point of the lipostat. Just in case you think the change in food preferences might be a conscious choice on their part, let me add a critical detail: Researchers see similar effects on appetite and food preferences in obese rats and mice who have undergone the same procedures.

Clearly, something about weight loss surgery alters how the brain nonconsciously regulates food intake and adiposity. And these effects result from changes in the jumble of connections and chemicals that the brain uses to do its job. So which changes, exactly, are responsible for the remarkable efficacy of certain weight-loss surgeries? No one knows for certain yet, but researchers like Randy Seeley of the University of Michigan and Hans-Rudi Berthoud of the Pennington Biomedical Research Center are closing in on the answer. And once we have that answer, it may empower us to manipulate the same circuits using tools that are less risky and less permanent than surgery, such as drugs, or, ideally, diet and lifestyle. We have much to learn about the brain, and within that unknown realm lies powerful tools for preventing and treating obesity.

At the same time, we already have quite a bit of information about how the brain works and which cues influence the nonconscious processes that drive overeating and fat gain. In the next chapter, I'll take what we've covered so far and see what practical value we can derive from it. The objective is to give the nonconscious brain the right cues, thereby aligning it with our conscious goals of leanness and health.

11

OUTSMARTING THE HUNGRY BRAIN

Throughout this book, we've seen the powerful influence of the nonconscious brain on our eating habits. Although these ancient circuits kept our ancestors alive and fertile in an environment of scarcity, they get us into trouble today by following the rules of a survival game that no longer exists. The same responses that once helped us thrive now drive us to overeat, gain fat, and develop diseases that kill us and sap our vigor. As Anthony Sclafani's and Eric Ravussin's cafeteria diet studies illustrate, the pull of a fattening food environment can cause us to overeat tremendously without any intention to do so, and without even awareness that it's happening. In today's cafeteria diet world, the deck is stacked against us.

But we also know that if we change our environment, we can change the cues we send to the nonconscious brain, aligning its motivations with our goals of leanness and health. If the cues your brain receives favor overeating, you'll probably overeat whether you intend to or not. If they don't favor overeating, you probably won't. If the nonconscious, impulsive brain is so influential, and if it acts according to the cues it receives, then the obvious path to effective weight management is to give it the right cues. How do we do that? We can divide approaches into two categories: things we can do as a nation and things we can do individually.

TACKLING THE OBESITY EPIDEMIC

There are a variety of public health measures that could guide our eating behavior in a healthier, more moderate direction as a nation. Yet if we want to have a meaningful impact on the nation's waistline, we're going to have to get a lot more serious about it than we currently are. Kevin Hall's research shows that Americans would have to reduce our average calorie intake by at least 218 Calories per day to return to our 1978 body weights, a reduction of nearly 10 percent. Alternatively, we'd have to burn 218 extra Calories exercising—which represents almost half an hour of jogging every day—and hope we don't work up an appetite. In reality, the situation is even more challenging, because not everyone overeats. If we *average* 218 excess Calories, what that really means is that some people don't overeat at all, while others overeat by 400 Calories or more per day. And the latter population is the one that public health measures would have to target. Humble arithmetic suggests that halfhearted efforts aren't going to cut it.

That said, what tools do we have at our disposal? The first strategy is simply to give people information on how to eat a healthy diet. Unfortunately, as we've seen, that approach by itself seems to have little impact on calorie intake. People do need good information to make good choices, but information alone isn't enough to substantially change behavior because it doesn't target the primary brain circuits that are in charge of calorie intake.

Nutrition labeling is an example of this approach. While labeling has had some victories like pressuring manufacturers into reducing the *trans* fat content of processed food,[155] telling people how many calories a food contains seems to have little impact on their overall calorie intake in real life. I would argue that it targets the wrong brain circuits.

Whether they view their approaches in these terms or not, public health professionals already have a number of weapons in their arsenal that target the impulsive brain. For example, our consumption patterns can be shifted by a strategy called *countermarketing*. This basically means running negative ads that associate certain products with bad feelings, unpleasant

---

155　As soon as the food industry was required to report *trans* fat on nutrition labels, the amount of it in US foods plummeted.

images, or disturbing information (an example of negative reinforcement). It's the opposite of what ads normally do. Antitobacco countermarketing, with its gory images of blackened lungs and tracheostomies, included ads on television and billboards, and stern warning labels on cigarette cartons. Coupled with crippling tobacco taxes and a growing number of public and private smoking bans, countermarketing contributed to the fact that Americans today smoke 70 percent fewer cigarettes per person than we did in 1963.

While the decline in cigarette smoking is a huge public health victory that has prevented a great deal of disease and suffering,[156] it was only possible due to a unique combination of factors that doesn't exist for fattening food. First of all, cigarettes aren't essential for life, whereas food is. This means we would have to countermarket against fattening foods but not others. It's much more challenging to split nutritional hairs than it is to oppose an addictive drug wholesale. Second, tobacco countermarketing was (and is) funded by tens of billions of dollars in legal settlements that resulted from antitobacco lawsuits in the 1990s. There is currently no such funding for food countermarketing efforts.

Another way to guide our eating behavior in the right direction is to make fattening foods a less attractive "deal" to the brain regions that guide our economic choices. Monetary cost is part of this equation, and as such it has received a lot of attention from public health professionals. Simply stated, we tend to buy less of specific foods as they become more expensive and more as they become cheaper. By making fattening food more expensive and slimming food cheaper, we can shift the cost-benefit balance to favor a more moderate calorie intake.

Several countries have already implemented taxes on foods that are considered unhealthy, including Denmark's tax on saturated-fat-rich foods and Mexico's tax on sugar-sweetened beverages. Yet fierce opposition from the food industry and the general public led Denmark to scrap its tax little more than a year after it was implemented. Mexico's soda tax has effectively, if modestly, reduced sugar-sweetened beverage intake in one of the fattest countries in the world. At the time of this writing, it's already vulnerable to new legislation only two years after being implemented. Not

---

156 It is probably the primary reason for our substantial and ongoing decline in lung cancer and heart attack rates.

surprisingly, the soda industry has been working hard to undermine it, and they have a lot of leverage since Mexico's government has strong ties to big soda (for example, Vicente Fox, the president of Mexico from 2000 to 2006, is the former president of Coca-Cola Mexico).

An alternative way to financially nudge our diet in the right direction is simply to change the way government subsidies are allocated to commodity crops, such as corn, soybeans, and wheat. These three food crops receive more subsidies than any others in the United States—totaling over $10 billion per year. They also happen to be the basis for many of our most fattening food ingredients, such as high-fructose corn syrup, white flour, soybean oil (the primary added fat in the United States), corn oil, and corn starch. In turn, the food industry uses these artificially cheap ingredients to compose extremely tempting and unbelievably low-cost foods—a deal that's hard for the brain to pass up. Essentially, taxpayers are subsidizing the very foods that make them fat and sick. Shifting subsidies away from foods that make us fat, and toward foods that don't, is a common-sense idea that could have a substantial impact on the American food system, and ultimately our waistlines. I'll leave it to the economists and farmers to understand the complex ramifications of changing our subsidy system, but it seems to me that we should be able to support farmers and food security without undermining our nation's health.

Convenience is an additional public health lever that can modify our eating behavior by impacting the brain circuits that govern economic choice. For example, in Detroit there's a zoning law that requires a minimum of five hundred feet between fast-food restaurants and schools (that's a bit less than two football fields). However, very few communities have passed such ordinances, and in most cases, the goal was to protect the character of a historic area rather than to promote health.

Food assistance programs like the Supplemental Nutrition Assistance Program (SNAP; formerly named the Food Stamp Program), which supports low-income American individuals and families, represent an additional intervention point. These programs target economically disadvantaged groups that are more likely to have obesity and chronic disease. Currently, 45 million Americans use SNAP, so it's a powerful leverage point for improving diet and reducing obesity risk. The government does limit what SNAP recipients can do with the money they receive: It can be used to buy groceries and seeds to grow food, but it cannot be used in restaurants. Although it's not difficult to compose an atrocious diet from grocery store foods, skipping restaurants is a step in the right direction. In the case of food assistance programs, we seem to be a little bit more comfortable with the government influencing food choices because the recipients aren't spending their own money. This could present a valuable opportunity to nudge our diet in the right direction, perhaps by providing financial incentives to buy healthy food rather than unhealthy food.

The food industry has long known that one of the most effective ways to get a person to buy a particular food is to show her an appealing image of it. If she has eaten the food before, or something similar, this visual cue triggers dopamine release and the motivation to eat the food (craving). This fundamental property of the human brain is one of the main reasons why the food industry spends tens of billions of dollars on advertising each year. The public health intervention here is obvious: Regulate food advertising. Yet in the United States, the government places few restrictions on food advertising. The research of the University of Connecticut's Rudd Center for Food Policy and Obesity shows that both children and adults in the United States are inundated every day by a deluge of food advertisements, most of which promote unhealthy, calorie-dense items. Americans are well aware of the insidious effect of food advertising on our food preferences and eating habits, yet we're reluctant to let the government step in.

In theory, the food industry can regulate itself, and this is a lot more palatable to the industry and the public than government regulation. All else being equal, food industry executives would rather not have their products fuel the obesity epidemic, and under pressure from government and private organizations, many companies have already taken action. One such effort is the Children's Food and Beverage Advertising Initiative (CFBAI), a voluntary program introduced in 2006 that regulates food advertising to US children under age twelve. Most of the largest food industry players are on board, including Kraft, Coca-Cola, and General Mills. CFBAI participants agree to restrict advertising of unhealthy foods on children's TV shows, requiring that advertised foods meet specific criteria for calorie, saturated fat, *trans* fat, sugar, and salt content. Foods don't have to be particularly healthy or slimming to meet the CFBAI's criteria, but the guidelines do restrict the worst offenders, and the food industry seems to comply with them reasonably well.

Not everyone has a rosy view of the CFBAI, however. The Rudd Center points out that children are often exposed to media that's targeted at adolescents twelve to eighteen years old, or adults, and these advertising arenas are unrestricted. So while these companies do restrict food advertising aimed directly at children, when they advertise unhealthy food to other age groups, there is a lot of collateral damage to children. The Rudd Center also reports that the nutrition standards used by the CFBAI are not as stringent as those recommended by independent organizations like the federal Interagency Working Group. Despite these problems, it's hard not to view the CFBAI as a step in the right direction.

Sadly, CFBAI is arguably the most impressive example of US food industry self-regulation in the interest of public health. The fundamental problem is that the food industry is governed by the incentives of a fiercely competitive free market economy, and these incentives sometimes conflict with human health. For example, in the 1970s, Nestlé initiated a campaign to promote its infant formula over breast milk in developing countries, despite the fact that contaminated local water supplies often made formula feeding dangerous. It was a profitable venture, but at the cost of substantial infant mortality. The situation was eventually resolved by boycotts and international regulation. Although not all instances of conflict between profits and human health are so egregious, competition nevertheless creates a "race to the bottom" in which companies jockey to produce the most compelling foods possible and advertise them in the most visible, attractive ways. To do this, they titillate the brain circuits that determine our innate nutrient and economic preferences—which are the same circuits that push us to overeat. The end result is a food system that's expertly, if unintentionally, crafted to drive overeating.

Reducing the fattening qualities of a food tends to make it less appealing to the average consumer, who then votes with his wallet by buying something else. For example, McDonald's has made several attempts to improve the health profile of its offerings, such as phasing out the Supersize option and adding fresh fruit and salads to its menu. Yet its US sales sagged in the wake of these changes. Other companies were quick to take up the slack, as illustrated by this quote from Hardee's CEO Andrew Puzder about their 1,400-Calorie "Monster Thickburger":

This is a burger for young hungry guys who want a really big, delicious, juicy, decadent burger. I hope our competitors keep promoting those healthy products, and we will keep promoting our big, juicy delicious burgers.

Hardee's sales were brisk following their introduction of the colossal hamburger.

Because many of the qualities that coax us to purchase a food are also those that make us overeat, any individual company that voluntarily makes its food less fattening is effectively entering the boxing ring with one hand tied behind its back.[157] Executives and shareholders both realize that having one hand tied behind your back usually leads to getting punched in the face. This creates a fundamental conflict between profits and public health—and in the grand scheme of things, profits usually prevail. In my view, the only realistic way out of this downward spiral is nationwide regulation that encourages a more slimming food environment while maintaining a level economic playing field for the food industry. This could happen through industry self-regulation and/or government regulation, but I seriously doubt industry has what it takes to self-regulate us out of our overeating problem.

Although public health legislation would undoubtedly be an effective antiobesity tool if we got serious about it, few Americans, and even fewer food industry executives, have much appetite for it. We tend to have a dim view of government meddling in general, and in particular when it comes to our food. I can sympathize with that perspective—we want to forge our own path through life. Yet we've already walked many miles down the wrong path, and it has caused millions of American children to grow up with obesity-related physical and metabolic disabilities before they have a choice.[158] These children will have a much higher chronic disease risk, a

---

157 There are niche markets for genuine health foods where this doesn't apply, but that's a small percentage of the total food market.

158 Two percent of American children are considered to have "extreme obesity" today, which is a dramatic increase over previous generations. These children have a greatly reduced capacity for ordinary childhood activities like running and climbing, and as adults they will have an exceptionally high risk of further disability, such as premature arthritis and diabetes. The prognosis is similar, if a bit less grim, for the additional 15 percent of children with less extreme obesity. I believe it's important to respect these children and their families rather than judging them, while still recognizing the seriousness of the problem and working together toward a solution.

lower life expectancy, and a lower quality of life as a result of improper nutrition. This tragic situation is largely preventable, yet our halfhearted attempts to do so have failed. It's time to ask ourselves a serious question: What do we care more about, the health of our nation's children or our freedom to be bombarded by cheap, fattening food?

SIX STEPS FOR A SLIMMING LIFESTYLE

For those of us who are motivated enough, we don't have to wait for legislation to tackle overeating. With the right information, we can craft a food environment and lifestyle that send slimming cues to the nonconscious brain, resulting in easier weight management. The objective is to create a situation in which the motivations of the conscious brain and the nonconscious brain are aligned—both working to support your goal of an optimal calorie intake. The following six steps translate the research I detailed throughout this book into practical steps you can take in your daily life.

1. Fix your food environment

Tempting food cues in your personal environment are powerful drivers of overeating due to their impact on brain areas that govern motivation and economic choice. Fortunately, one of the most effective tools in our arsenal is also one of the simplest: Reduce your exposure to food cues. Here are three measures you can take to do so.

First, get rid of all tempting, calorie-dense foods that are easy to grab and eat in your home and work environment—particularly those that are readily visible on counters and tables. This includes things like chips and cookies but also some relatively healthy items like salted nuts. Get rid of the ice cream in your freezer. Don't give yourself the option to eat these foods, and you'll find that you crave them less.

Second, reduce your exposure to food cues in general. It's possible to overeat even healthy foods, so don't tempt yourself too much. Limit the amount of visible food in your personal food environment at home and at work, particularly snack foods that are easy to grab and eat. Minimize your exposure to food advertising on television and elsewhere if you can.

Third, create effort barriers to eating. These barriers don't have to be large to be effective. For example, if you have to peel an orange to eat it, you probably won't go for it unless you're genuinely hungry. The same goes for nuts in their shells. Perhaps the most stringent effort barrier is to limit the food in your kitchen to items you'd have to cook or reheat to eat. Chances are, if you have to cook something, you won't eat it between meals unless you really need it.

Putting this together, a healthy food environment is one that effortlessly guides your eating behavior in the right direction. It doesn't contain tempting, calorie-dense foods or ads that remind you of those foods; it contains little visible food in general; and it provides small effort barriers to eating the healthy food that is visible. Imagine a kitchen in which the only visible foods are whole fruit and nuts in shells, and eating anything else would require reheating something from the refrigerator. Now imagine a workplace in which the only available food is in a refrigerator, with no visible food on tables or countertops. That's the goal.

2. Manage your appetite

If your brain thinks you're starving, it will eventually wear you down, no matter how strong your resolve. The solution is to give it the cues it needs to realize you aren't starving.

The most straightforward way to do this is to choose foods that send strong satiety signals to the brain stem but contain a moderate number of calories. These are foods that have a lower calorie density, higher protein and/or fiber content, and a moderate level of palatability. This tends to include simple foods that are closer to their natural state, such as fresh fruit, vegetables, potatoes, fresh meats, seafood, eggs, yogurt, whole grains, beans, and lentils. Bread is surprisingly calorie-dense, even when it's made from whole grains, so it can be easy to overeat. It may be preferable to get your starch from water-rich foods like potatoes, sweet potatoes, beans, and oatmeal than flour-based foods like bread and crackers. And foods based on white flour in particular, which tend to have a high calorie density and low fiber content, are definitely off the menu.

In the long run, you also need to keep the lipostat comfortable with your target weight. We don't know as much about how to satisfy the lipostat, and we'll need more research to get a clearer view, but there is suggestive evidence that eating more protein and limiting highly rewarding foods can help. Regular physical activity, restorative sleep, and stress management may also support a leaner adiposity set point, facilitating weight loss and maintenance.

3. Beware of food reward

The brain values foods that contain calorie-dense combinations of fat, sugar, starch, protein, salt, and other elements, and it sets your motivation to eat those foods accordingly. This motivation is partially independent of hunger, such that it's easy to blow past satiety signals if you're eating something you love—think ice cream, brownies, french fries, chocolate, and bacon. These foods are a lot more rewarding than anything our distant ancestors ate, and they can powerfully drive cravings, overeating, and eventually, deeply ingrained unhealthy eating habits.

When we eat simple foods that are less dense in calories and closer to their natural states, they're still enjoyable but they don't have that intensely rewarding edge that drives us to overdo it. These include things like fruit, vegetables, potatoes, beans, oatmeal, eggs, plain yogurt, fresh meat, and seafood. Nuts may not be an ideal diet food, but they are less calorie-dense than they might seem because some of their calories pass through the digestive system unabsorbed. Choosing unsalted nuts reduces their reward value to a reasonable level. Simple foods like these help the satiety system and the lipostat stay in control of our eating behavior, matching our calorie intake to our true needs.

Not everyone finds the same foods highly rewarding, but most people have a pretty good idea what their own problem foods are. Common examples include chocolate, pizza, potato chips, tortilla chips, french fries, cookies, cake, and ice cream. Keep these foods out of your personal food environment. You can still eat them occasionally.

If you're a dessert fiend and only decadence will do, try eating a piece of fruit with your dinner. This will promote sensory-specific satiety and reduce your craving for sweet foods at the end of the meal.

Watch out for habit-forming drugs, such as alcohol, caffeine, and theobromine (found in chocolate). These are inherently rewarding and can motivate us to take in calories we don't need, such as those in beer, cream and sugar, chocolate, and soda. When you consider that a single alcoholic beverage contains 90 to 180 calories, a can of soda contains about 140 calories, coffee beverages can contain up to 500 calories—and we don't drink any of those out of hunger—it's not hard to understand how they can contribute to excess adiposity. It's best to favor caffeinated beverages that don't contain calories, such as green tea and black coffee; avoid soda; and if you drink alcohol, consider limiting your intake to one lower-calorie beverage, such as wine or spirits, per day.

4. Make sleep a priority

I hope I've already dispelled the myth that sleep is a waste of time. Restorative sleep is an important cue for the nonconscious brain that has a major impact on performance and eating behavior—even if we aren't directly aware of it.

The first step toward restorative sleep is simply to spend enough time in bed. This may be all it takes for many people to be well rested. For people who have trouble sleeping, it helps to make sure your room is completely dark at night, allow your room to cool down in the evening if possible, and only use your bed for sleeping and sex.

Your circadian rhythm is a related cue that affects your sleep quality and eating behavior via largely nonconscious processes. To give your circadian rhythm the right cues, try to get into bed and wake up at about the same time each day. Make sure to get bright, blue-spectrum light in the morning or at midday, ideally by spending time outside. And in the evening, avoid bright, blue-spectrum light by replacing full-spectrum bulbs with warm white bulbs, dimming lights, using programs such as f.lux on your electronic devices, and/or wearing blue-blocking glasses.

If you have a more serious condition that undermines sleep quality, such as sleep apnea, seek professional treatment. Most sleep apnea is readily treatable, and doing so can substantially improve your health, performance, and quality of life.

5. Move your body

Regular physical activity can help manage your appetite and weight in at least two ways. First, it increases the number of calories you use, making it less likely you'll overeat. Studies show that when people with excess weight exercise regularly, their calorie intake tends to go up, but usually not enough to compensate for the calories they burn (although this does vary by individual). Second, physical activity may also help maintain the lipostat in the brain, encouraging a naturally lower level of adiposity in the long run.

Our distant ancestors had a word for exercise: *life*. Movement has always been a key part of our species' daily activity, and our bodies require it to function properly. It's a fundamental ingredient for good health, physical and cognitive performance, emotional health, and healthy aging. As such, it's an essential part of a healthy lifestyle, whether or not your goal is weight management.

The most important thing to remember about physical activity is simply to do it, every day if possible. Whether you're walking, gardening, playing tennis, riding a bicycle, or strength training, it all counts. However, the ideal situation is to train your body using a mixture of different types of activity, as suggested by the US Department of Health and Human Services (HHS). HHS recommends a combination of moderate-intensity activities, such as brisk walking, high-intensity cardiovascular activities like running, and strength-building activities like lifting weights.

It's important to choose activities that fit into your schedule and that you enjoy; otherwise, they may not be sustainable. I think commuting on foot or by bicycle is a great way to build physical activity into your daily life in a time-efficient way. It may not be an option for everyone, but it is for many more people than may realize it. If your commute is too long to do by bike, try parking a few miles away from work and walking, jogging, or biking the rest of the way. You might find that you enjoy it. Sports like basketball and tennis are a fun, social way to be active. In many areas, municipal leagues and public tennis courts make this an easy and inexpensive option.

6. Manage stress

The threat response system evolved to protect us, but sometimes in the modern world it can undermine our quality of life and our best intentions to eat the right amount of food. I'll outline five actions you can take to identify the problem and manage stress eating by giving your threat response system the right cues. The first action is simply to identify whether or not you're a stress eater. If you are, then you probably already know it. The second action is to identify the stressor(s)—particularly chronic stressors you don't feel you can control. These often include work stress, money, health problems, prolonged caregiving, interpersonal conflict, and/or a lack of social support.

The third action is to try to mitigate the stressor. There are multiple ways to do this. Can you fix it or avoid it? If not, is there a way you can turn what seems like an uncontrollable stressor into a controllable stressor? For example, if you're stressed about money, can you make a concrete plan to improve your finances? If you have stressful health problems, can you lay out a concrete road map for managing your condition as effectively as possible? Making a plan will probably help you mitigate your stressor, but even if it doesn't, it gives you a feeling of control that may reduce your drive to stress eat.

Another way to mitigate stress is by practicing mindfulness meditation. Mindfulness is a state of intentional, nonjudgmental awareness of the present moment, and meditation is an effective way to cultivate it. Most of what stresses us has little to do with what's happening right now—it's usually about what *might* happen in the future. I might not make that work deadline. I might develop diabetes. My partner might leave me. I might not make my credit card payment. Sometimes these concerns are rational and deserve attention, but they often run away with our minds and emotions in a way that isn't helpful. By training ourselves to focus on the present moment, we can guide our thoughts in a more constructive direction.

There are many ways to meditate, but here's a simple technique that works:

Find a comfortable seat where you can maintain a straight but relaxed spine. Keep your eyes open and your gaze slightly downward. Then just pay attention to the rise and fall of your abdomen as you breathe. You'll notice things happening around you, and your mind will wander, but just keep gently bringing your attention back to your breath. Start with five minutes, and work your way up to fifteen. Think of it like exercise; it's hard when you're out of shape, but it gets easier the more you practice. Although research on mindfulness training has a long way to go, there is already substantial evidence that it can reduce stress and increase quality of life, and limited evidence that it can also improve health.

The fourth action is to replace stress eating with more constructive coping methods. Are there other enjoyable things you can substitute for comfort food when you're stressed? How about calling a friend, making love, reading a good book, jogging, taking a hot bath, or gardening?

At the risk of sounding like a broken record, the fifth action is to remove calorie-dense comfort food from your personal surroundings at home and at work. In the absence of highly rewarding items, there's less of an incentive to self-medicate your stress with food.

The strategies I've outlined above should help you manage your weight by providing the right cues to your nonconscious brain so that it's aligned with your conscious goal of being lean and healthy. If weight control is your goal, I encourage you to think about how you can use the six steps I've outlined as efficiently and sustainably as possible. Which steps will deliver you the most benefit, and which the least?

For example, I'm very sensitive to highly rewarding foods in my immediate surroundings. I tend to overdo it when french fries, cookies, chips, or other tempting foods are easy to grab—especially when I've had a drink of alcohol. I also eat too much when my food has a low satiety value per calorie. For me, it makes sense to focus on maintaining a supportive food environment and eating simple foods that have a lower calorie density and a higher satiety value. These are the approaches that give me the most payback for my effort. On the other hand, I don't overeat when I'm stressed, I already sleep well, and I get plenty of physical activity. These factors aren't a high priority for me. It's important to identify your individual priorities and use them to craft a weight and health management plan that works for you.

I hope that by understanding the causes of overeating using scientific research, developing more effective individual and public health strategies, and having the courage to use them, together we can rein in our hungry brains.

Acknowledgments

Although I was the one who manned the keyboard, this book is the work of many remarkable people. First and foremost, I'd like to thank the countless researchers whose dedication to understanding the natural world made this possible. I'm particularly indebted to those who made time to explain their work to me, including Anthony Sclafani, Brian Wood, Bruce Winterhalder, Camillo Padoa-Schioppa, Cliff Saper, Dan Pardi, Deanna Arble, Elissa Epel, Ellen Schur, Eric Ravussin, Harvey Grill, Herman Pontzer, Josh Thaler, Kent Berridge, Kevin Gurney, Kevin Hall, Kim Hill, Leann Birch, Leonard Epstein, Leslie Lieberman, Marcus Stephenson-Jones, Marie-Pierre St-Onge, Mark Wilson, Mary Dallman, Mike Schwartz, Mike Shadlen, Peter Redgrave, Richard Palmiter, Ross McDevitt, Roy Wise, Rudy Leibel, Ruth Harris, Sadaf Farooqi, Staffan Lindeberg, and Yvonne Ulrich-Lai. I hope I've done justice to their research. In addition, Brian Wood, Camillo Padoa-Schioppa, Dan Pardi, Ellen Schur, Kevin Hall, Leonard Epstein, Marcus Stephenson-Jones, Mark Wilson, Mike Schwartz, Peter Redgrave, and Staffan Lindeberg provided valuable scientific feedback on my chapter drafts. Kevin Hall generously provided raw data for figure 3. Peter Redgrave patiently answered my numerous questions as I labored to grasp his hypothesis about the role of the basal ganglia in action selection. Jeremy Landen helped me research historical US sweetener intake. Ashley Mason convinced me to write a chapter on stress. Ross McDevitt provided a wonderful figure illustrating locomotion in mice when they're high on cocaine. Brian Wood provided lovely photos of the Hadza collecting and preparing food. Ellen

Schur, Susan Melhorn, Mary K. Askren, and the University of Washington Diagnostic Imaging Sciences Center generously agreed to scan my brain while I looked at images of junk food. Susan in particular put substantial work into preparing my fMRI images and tackling the difficult task of making them compatible with grayscale print.

I'd also like to thank my agent, Howard Yoon, and my editor at Flatiron Books, Whitney Frick, for taking a chance on a first-time author. Kristin Mehus-Roe provided helpful editorial feedback on my proposal and connected me with Howard. Janine Jagger, Beth Sosik, and Zen Wolfang provided useful comments on specific chapters. Rachel Holtzman worked with me extensively on the complete draft to help make the book as engaging and accessible as possible. Shizuka Aoki provided the wonderful illustrations and gave my graphs and schematics a much-needed makeover.

Finally, a heartfelt thanks to Mike Schwartz and all the lovely people I met while I was working with him at the University of Washington. Without the inspiration and perspiration of that experience, this book wouldn't exist.

Notes

1   *Dietary Guidelines for Americans:* "Nutrition and Your Health," *Dietary Guidelines for Americans*, US Department of Agriculture and US Department of Health and Human Services (1980).

2   *rate more than doubled:* C. D. Fryar, M. D. Carroll, and C. L. Ogden, "Prevalence of Overweight, Obesity, and Extreme Obesity Among Adults: United States, Trends 1960–1962 through 2007–2008," National Center for Health Statistics (2012), 1–6.

2   *refined starch and sugar:* G. Taubes, *Good Calories, Bad Calories: Fats, Carbs, and the Controversial Science of Diet and Health*, reprint edition (New York: Anchor, 2008), 640; N. Teicholz, *The Big Fat Surprise: Why Butter, Meat & Cheese Belong in a Healthy Diet* (New York: Simon & Schuster, 2015), 496.

2n *USDA and CDC data:* G. L. Austin, L. G. Ogden, and J. O. Hill, "Trends in Carbohydrate, Fat, and Protein Intakes and Association with Energy Intake in Normal-Weight, Overweight, and Obese Individuals: 1971–2006," *American Journal of Clinical Nutrition* 93, no. 4 (April 2011): 836–43; USDA, "USDA Economic Research Service—Food Availability (Per Capita) Data System," 2013, cited October 31, 2013, http://www.ers.usda.gov/data-products/food-availability -(per-capita)-data-system.aspx.

2   *recommended limiting fat intake:* W. O. Atwater, "Foods: Nutritive Value and Cost," USDA, *Farmers Bulletin,* no. 23 (1894).

2   *Americans who don't:* X. Guo, B. A. Warden, S. Paeratakul, and G. A. Bray, "Healthy Eating Index and Obesity," *European Journal of Clinical Nutrition* 58, no. 12 (December 2004): 1580–86; P. A. Quatromoni, M. Pencina, M. R. Cobain, P. F. Jacques, and R. B. D'Agostino, "Dietary Quality Predicts Adult Weight Gain: Findings from the Framingham Offspring Study," *Obesity* 14, no. 8 (August 2006): 1383–91.

2n *metabolic health, and cardiovascular disease:* M. Kratz, T. Baars, and S. Guyenet, "The Relationship Between High-Fat Dairy Consumption and Obesity, Cardio-vascular, and Metabolic Disease," *European Journal of Nutrition* 52, no. 1 (February 2013): 1–24.

3   *our calorie intake:* USDA, "USDA Economic Research Service—Food Availability
     (Per Capita)."

4   *do precisely that:* Fryar, Carroll, and Ogden, "Prevalence of Overweight, Obesity."

5   *in the affluent world:* D. Lieberman, *The Story of the Human Body: Evolution,
     Health, and Disease* (New York: Pantheon Books, 2013); S. Lindeberg, *Food and
     Western Disease: Health and Nutrition from an Evolutionary Perspective,* 1st ed.
     (Hoboken, NJ: Wiley-Blackwell, 2010), 368.

1. The Fattest Man on the Island

7   *touched by industrialization:* Lindeberg, *Food and Western Disease*; S. Lindeberg
     and B. Lundh, "Apparent Absence of Stroke and Ischaemic Heart Disease in a
     Traditional Melanesian Island: A Clinical Study in Kitava," *Journal of Internal
     Medicine* 233, no. 3 (March 1993): 269–75; S. Lindeberg, E. Berntorp, P. Nilsson-
     Ehle, A. Terént, and B. Vessby, "Age Relations of Cardiovascular Risk Factors in
     a Traditional Melanesian Society: The Kitava Study," *American Journal of Clinical
     Nutrition* 66, no. 4 (October 1997): 845–52.

7   *even in old age:* Ibid.

7   *ancestors might have lived:* Lindeberg, *Food and Western Disease;* P. F. Sinnett
     and H. M. Whyte, "Epidemiological Studies in a Total Highland Population,
     Tukisenta, New Guinea: Cardiovascular Disease and Relevant Clinical, Electro-
     cardiographic, Radiological, and Biochemical Findings," *Journal of Chronic Diseases*
     26, no. 5 (May 1973): 265–90; K. T. Lee, R. Nail, L. A. Sherman, M. Milano,
     C. Deden, H. Imai, et al., "Geographic Pathology of Myocardial Infarction," *American
     Journal of Cardiology* 13 (January 1964): 30–40; H. C. Trowell and D. P. Burkitt,
     Western Diseases: Their Emergence and Prevention (Cambridge, MA: Harvard
     University Press, 1981), 474; F. W. Marlowe and J. C. Berbesque, "Tubers as
     Fallback Foods and Their Impact on Hadza Hunter-Gatherers," *American Journal
     of Physical Anthropology* 140, no. 4 (December 2009): 751–58; T. Teuscher, J. B.
     Rosman, P. Baillod, and A. Teuscher, "Absence of Diabetes in a Rural West
     African Population with a High Carbohydrate/Cassava Diet," *Lancet* 329, no. 8536
     (April 1987): 765–68; R. B. Lee, *The !Kung San: Men, Women, and Work in a
     Foraging Society* (Cambridge, UK: Cambridge University Press, 1979).

8   *Kitavan examined by Lindeberg:* Lindeberg, *Food and Western Disease.*

8   *obesity and chronic disease:* Trowell and Burkitt, *Western Diseases: Their Emergence.*

8   *jobs involved manual labor:* National Bureau of Economic Research Economic
     History Association, *Output, Employment, and Productivity in the United States After
     1800* (National Bureau of Economic Research, 1966), 684.

9   *body mass index (BMI):* L. A. Helmchen and R. M. Henderson, "Changes in the
     Distribution of Body Mass Index of White US Men, 1890–2000," *Annals of
     Human Biology* 31, no. 2 (April 2004): 174–81.

9   *common as it is today:* J. N. Wilford, "Tooth May Have Solved Mummy Mystery,"
     New York Times, June 27, 2007, cited March 11, 2016, http://www.nytimes.com/2007
     /06/27/world/middleeast/27mummy.html.

9   *one out of three:* Fryar, Carroll, and Ogden, "Prevalence of Overweight, Obesity."

9   *one out of 17*: Ibid.

9   *increased nearly fivefold:* C. L. Ogden and M. D. Carroll, "Prevalence of Obesity
        Among Children and Adolescents: United States, Trends 1963–1965 through
        2007–2008," 2013, cited October 31, 2013, http://www.cdc.gov/nchs/data/hestat
        /obesity_adult_09_10/obesity_adult_09_10.htm.

11   *linked to excess weight:* A. Stokes, "Using Maximum Weight to Redefine Body Mass
        Index Categories in Studies of the Mortality Risks of Obesity," *Population Health
        Metrics* 12, no. 1 (2014): 6; A. Stokes and S. H. Preston, "Revealing the Burden of
        Obesity Using Weight Histories," *Proceedings of the National Academy of Sciences
        of the United States of America* 113, no. 3 (January 2016): 572–77.

11   *rerouted to lose weight:* J. Ponce, "New Procedure Estimates for Bariatric Surgery:
        What the Numbers Reveal," Connect, May 2014, cited March 11, 2016, http://
        connect.asmbs.org/may-2014-bariatric-surgery-growth.html.

11   *contained in chemical bonds:* J. L. Hargrove, "History of the Calorie in Nutrition,"
        Journal of Nutrition 136, no. 12 (December 2006): 2957–61.

11   *furnace of the human body:* W. O. Atwater, "The Potential Energy of Food: The
        Chemistry and Economy of Food, III," *Century* 34 (1887): 397–405.

12   *convention begun by Atwater:* Ibid.

12   *energy leaving the body:* W. O. Atwater, *Experiments on the Metabolism of Matter and
        Energy in the Human Body, 1898–1900* (US Government Printing Office, 1902), 166.

13   *human furnace is concerned:* Ibid.

13   *weight maintenance, or weight gain:* O. Lammert, N. Grunnet, P. Faber, K. S.
        Bjørnsbo, J. Dich, L. O. Larsen, et al., "Effects of Isoenergetic Overfeeding of
        Either Carbohydrate or Fat in Young Men," *British Journal of Nutrition* 84, no. 2
        (2000): 233–45; T. J. Horton, H. Drougas, A. Brachey, G. W. Reed, J. C. Peters,
        and J. O. Hill, "Fat and Carbohydrate Overfeeding in Humans: Different Effects
        on Energy Storage," *American Journal of Clinical Nutrition* 62, no. 1 (July 1995):
        19–29; R. L. Leibel, J. Hirsch, B. E. Appel, and G. C. Checani, "Energy Intake
        Required to Maintain Body Weight Is Not Affected by Wide Variation in Diet
        Composition," *American Journal of Clinical Nutrition* 55, no. 2 (February 1992):
        350–55; N. Grey and D. M. Kipnis, "Effect of Diet Composition on the Hyperin-
        sulinemia of Obesity," *New England Journal of Medicine* 285, no. 15 (October 7,
        1971): 827–31; C. Bogardus, B. M. LaGrange, E. S. Horton, and E. A. Sims,
        "Comparison of Carbohydrate-Containing and Carbohydrate-Restricted
        Hypocaloric Diets in the Treatment of Obesity: Endurance and Metabolic Fuel
        Homeostasis During Strenuous Exercise," *Journal of Clinical Investigation* 68,
        no. 2 (August 1981): 399–404; P. M. Piatti, F. Monti, I. Fermo, L. Baruffaldi, R.
        Nasser, G. Santambrogio, et al., "Hypocaloric High-Protein Diet Improves
        Glucose Oxidation and Spares Lean Body Mass: Comparison to Hypocaloric
        High-Carbohydrate Diet," *Metabolism* 43, no. 12 (December 1994): 1481–87; A.
        Golay, A. F. Allaz, Y. Morel, N. de Tonnac, S. Tankova, and G. Reaven, "Similar
        Weight Loss with Low- or High-Carbohydrate Diets," *American Journal of
        Clinical Nutrition* 63, no. 2 (February 1996): 174–78.

13n   *meaningful differences in adiposity:* C. B. Ebbeling, J. F. Swain, H. A. Feldman,
        W. W. Wong, D. L. Hachey, E. Garcia-Lago, et al., "Effects of Dietary Composition

on Energy Expenditure During Weight-Loss Maintenance," *Journal of the American Medical Association* 307, no. 24 (June 27, 2012): 2627–34; K. D. Hall, T. Bemis, R. Brychta, K. Y. Chen, A. Courville, E. J. Crayner, et al., "Calorie for Calorie, Dietary Fat Restriction Results in More Body Fat Loss than Carbohydrate Restriction in People with Obesity," *Cell Metabolism* 22, no. 3 (September 2015): 427–36.

14   *calories are left per person:* USDA, "USDA Economic Research Service—Food Availability (Per Capita)."

14   *tally up the calories:* E. S. Ford and W. H. Dietz, "Trends in Energy Intake Among Adults in the United States: Findings from NHANES," *American Journal of Clinical Nutrition* 97, no. 4 (April 2013): 848–53.

14   *increase in weight:* K. D. Hall, J. Guo, M. Dore, C. C. Chow, "The Progressive Increase of Food Waste in America and Its Environmental Impact," *PLOS ONE* 4, no. 11 (November 2009): e7940.

14   *same period of time:* B. Swinburn, G. Sacks, and E. Ravussin, "Increased Food Energy Supply Is More Than Sufficient to Explain the US Epidemic of Obesity," *American Journal of Clinical Nutrition* 90, no. 6 (December 2009): 1453–56.

15   *weight gain and weight loss:* K. D. Hall, G. Sacks, D. Chandramohan, C. C. Chow, Y. C. Wang, S. L. Gortmaker, et al., "Quantification of the Effect of Energy Imbalance on Bodyweight," *Lancet* 378, no. 9793 (August 27, 2011): 826–37.

16   *every pound you want to lose:* Ibid.

16   *happens in later chapters:* Ibid.

16n *fantastic paper on this:* Hall, Guo, Dore, and Chow, "The Progressive Increase of Food Waste."

17   *shown in figure 5:* S. C. Davis and S. W. Diegel, *Transportation Energy Data Book: Edition 32* (US Department of Energy, 2013).

19   *milk chocolate, and peanut butter:* A. Sclafani and D. Springer, "Dietary Obesity in Adult Rats: Similarities to Hypothalamic and Human Obesity Syndromes," *Physiology and Behavior* 17, no. 3 (September 1976): 461–71.

19   *high in fat and/or sugar:* B. P. Sampey, A. M. Vanhoose, H. M. Winfield, A. J. Freemerman, M. J. Muehlbauer, P. T. Fueger, et al., "Cafeteria Diet Is a Robust Model of Human Metabolic Syndrome with Liver and Adipose Inflammation: Comparison to High-Fat Diet," *Obesity* (Silver Spring, MD) 19, no. 6 (June 2011): 1109–17.

20   *calorie intake in obesity:* D. M. Dreon, B. Frey-Hewitt, N. Ellsworth, P. T. Williams, R. B. Terry, and P. D. Wood, "Dietary Fat: Carbohydrate Ratio and Obesity in Middle-Aged Men," *American Journal of Clinical Nutrition* 47, no. 6 (June 1, 1988): 995–1000; D. Kromhout, "Energy and Macronutrient Intake in Lean and Obese Middle-Aged Men (The Zutphen Study)," *American Journal of Clinical Nutrition* 37, no. 2 (February 1983): 295–99; W. C. Miller, M. G. Niederpruem, J. P. Wallace, and A. K. Lindeman, "Dietary Fat, Sugar, and Fiber Predict Body Fat Content," *Journal of the American Dietetic Association* 94, no. 6 (June 1994): 612–15.

20   *measuring calorie intake:* S. W. Lichtman, K. Pisarska, E. R. Berman, M. Pestone, H. Dowling, E. Offenbacher, et al., "Discrepancy Between Self-Reported and

Actual Caloric Intake and Exercise in Obese Subjects," *New England Journal of Medicine* 327, no. 27 (December 31, 1992): 1893–98; E. Ravussin, S. Lillioja, T. E. Anderson, L. Christin, and C. Bogardus, "Determinants of 24-Hour Energy Expenditure in Man: Methods and Results Using a Respiratory Chamber," *Journal of Clinical Investigation* 78, no. 6 (December 1986): 1568–78; L. G. Bandini, D. A. Schoeller, H. N. Cyr, and W. H. Dietz, "Validity of Reported Energy Intake in Obese and Nonobese Adolescents," *American Journal of Clinical Nutrition* 52, no. 3 (September 1, 1990): 421–5; E. Ravussin, B. Burnand, Y. Schutz, and E. Jéquier, "Twenty-Four-Hour Energy Expenditure and Resting Metabolic Rate in Obese, Moderately Obese, and Control Subjects," *American Journal of Clinical Nutrition* 35, no. 3 (March 1, 1982): 566–73.

20   *how much, they eat:* Lichtman, Pisarska, Berman, Pestone, Dowling, Offenbacher, et al., "Discrepancy Between Self-Reported and Actual"; Bandini, Schoeller, Cyr, and Dietz, "Validity of Reported Energy Intake"; J. O. Fisher, R. K. Johnson, C. Lindquist, L. L. Birch, and M. I. Goran, "Influence of Body Composition on the Accuracy of Reported Energy Intake in Children," *Obesity Research* 8, no. 8 (November 1, 2000): 597–603.

21   *entrées, snacks, and beverages:* R. Rising, S. Alger, V. Boyce, H. Seagle, R. Ferraro, A. M. Fontvieille, et al., "Food Intake Measured by an Automated Food-Selection System: Relationship to Energy Expenditure," *American Journal of Clinical Nutrition* 55, no. 2 (February 1, 1992): 343–49.

21   *"human cafeteria diet" studies:* D. E. Larson, P. A. Tataranni, R. T. Ferraro, and E. Ravussin, "Ad Libitum Food Intake on a 'Cafeteria Diet' in Native American Women: Relations with Body Composition and 24-H Energy Expenditure," *American Journal of Clinical Nutrition* 62, no. 5 (November 1, 1995): 911–17; D. Larson, R. Rising, R. Ferraro, and E. Ravussin, "Spontaneous Overfeeding with a 'Cafeteria Diet' in Men: Effects on 24-Hour Energy Expenditure and Substrate Oxidation," *International Journal of Obesity and Related Metabolic Disorders* 19, no. 5 (May 1995): 331–37.

2. The Selection Problem

23   *a fin on land:* S. Kumar and S. B. Hedges, "A Molecular Timescale for Vertebrate Evolution," *Nature* 392, no. 6679 (April 30, 1998): 917–20.

23   *human decision-making apparatus:* M. Stephenson-Jones, E. Samuelsson, J. Ericsson, B. Robertson, and S. Grillner, "Evolutionary Conservation of the Basal Ganglia as a Common Vertebrate Mechanism for Action Selection," *Current Biology* 21, no. 13 (July 12, 2011): 1081–91.

24   *one paint nozzle:* P. Redgrave, T. J. Prescott, and K. Gurney, "The Basal Ganglia: A Vertebrate Solution to the Selection Problem?" *Neuroscience* 89, no. 4 (1999): 1009–23.

24n   *the brains of flies:* N. J. Strausfeld and F. Hirth, "Deep Homology of Arthropod Central Complex and Vertebrate Basal Ganglia," *Science* 340, no. 6129 (April 12, 2013): 157–61; V. G. Fiore, R. J. Dolan, N. J. Strausfeld, and F. Hirth, "Evolutionarily Conserved Mechanisms for the Selection and Maintenance of Behavioural

Activity," *Philosophical Transactions of the Royal Society of London, Series B: Biological Sciences* 370, no. 1684 (December 2015).

25 *in a living organism:* Ibid.

26 *called the basal ganglia:* Ibid.

26 *parts of the brain:* Stephenson-Jones, Samuelsson, Ericsson, Robertson, and Grillner, "Evolutionary Conservation of the Basal Ganglia"; Fiore, Dolan, Strausfeld, and Hirth, "Evolutionarily Conserved Mechanisms."

26 *from the basal ganglia:* S. Grillner, J. Hellgren, A. Ménard, K. Saitoh, and M. A. Wikström, "Mechanisms for Selection of Basic Motor Programs—Roles for the Striatum and Pallidum," *Trends in Neurosciences* 28, no. 7 (July 2005): 364–70; A. Ménard and S. Grillner, "Diencephalic Locomotor Region in the Lamprey— Afferents and Efferent Control," *Journal of Neurophysiology* 100, no. 3 (September 2008): 1343–53.

27 *involved in planning behavior:* Fiore, Dolan, Strausfeld, and Hirth, "Evolutionarily Conserved Mechanisms"; F. M. Ocaña, S. M. Suryanarayana, K. Saitoh, A. A. Kardamakis, L. Capantini, B. Robertson, et al., "The Lamprey Pallium Provides a Blueprint of the Mammalian Motor Projections from Cortex," *Current Biology* 25, no. 4 (February 16, 2015): 413–23.

27 *to track its prey:* Ibid.

27 *the striatum (figure 7):* Fiore, Dolan, Strausfeld, and Hirth, "Evolutionarily Conserved Mechanisms."

28 *select the strongest bid:* Redgrave, Prescott, and Gurney, "The Basal Ganglia."

28 *particular action (figure 7):* Ocaña, Suryanarayana, Saitoh, Kardamakis, Capantini, Robertson, et al., "The Lamprey Pallium"; J. G. McHaffie, T. R. Stanford, B. E. Stein, V. Coizet, and P. Redgrave, "Subcortical Loops Through the Basal Ganglia," *Trends in Neurosciences* 28, no. 8 (August 2005): 401–407.

28n *decisions under complex conditions:* R. Bogacz and K. Gurney, "The Basal Ganglia and Cortex Implement Optimal Decision Making Between Alternative Actions," *Neural Computation* 19, no. 2 (January 5, 2007): 442–77.

29 *our body weight:* K. N. Frayn, *Metabolic Regulation: A Human Perspective* (Chichester, UK: Wiley-Blackwell, 2010).

30 *lampreys and mammals:* Stephenson-Jones, Samuelsson, Ericsson, Robertson, and Grillner, "Evolutionary Conservation of the Basal Ganglia."

30 *same chemical messengers:* Ibid.

30 *some 560 million years ago:* S. Grillner, B. Robertson, and M. Stephenson-Jones, "The Evolutionary Origin of the Vertebrate Basal Ganglia and Its Role in Action Selection," *Journal of Physiology* 591, no. 22 (November 15, 2013): 5425–31.

31 *process called exaptation:* Stephenson-Jones, Samuelsson, Ericsson, Robertson, and Grillner, "Evolutionary Conservation of the Basal Ganglia."

31n *learn simple tasks:* I. Q. Whishaw and B. Kolb, "The Mating Movements of Male Decorticate Rats: Evidence for Subcortically Generated Movements by the Male but Regulation of Approaches by the Female," *Behavioural Brain Research* 17, no. 3 (October 1985): 171–91; D. A. Oakley, "Performance of Decorticated Rats in a

Two-Choice Visual Discrimination Apparatus," *Behavioural Brain Research* 3, no. 1 (July 1981): 55–69.

32 *like the lamprey pallium:* G. E. Alexander, M. E. DeLong, and P. L. Strick, "Parallel Organization of Functionally Segregated Circuits Linking Basal Ganglia and Cortex," *Annual Review of Neuroscience* 9 (1986): 357–81; F. A. Middleton and P. L. Strick, "Basal Ganglia and Cerebellar Loops: Motor and Cognitive Circuits," *Brain Research Reviews* 31, nos. 2,3 (March 2000): 236–50.

32 *numerous other processes:* McHaffie, Stanford, Stein, Coizet, and Redgrave, "Subcortical Loops"; Middleton and Strick, "Basal Ganglia and Cerebellar Loops"; P. Romanelli, V. Esposito, D. W. Schaal, and G. Heit, "Somatotopy in the Basal Ganglia: Experimental and Clinical Evidence for Segregated Sensorimotor Channels," *Brain Research Reviews* 48, no. 1 (February 2005): 112–28.

33 *competing motivations and emotions:* Fiore, Dolan, Strausfeld, and Hirth, "Evolutionarily Conserved Mechanisms"; A. E. Kelley, "Ventral Striatal Control of Appetitive Motivation: Role in Ingestive Behavior and Reward-Related Learning," *Neuroscience and Biobehaviorial Reviews* 27, no. 8 (January 2004): 765–76; A. E. Kelley, "Neural Integrative Activities of Nucleus Accumbens Subregions in Relation to Learning and Motivation," *Psychobiology* 27, no. 2 (June 1, 1999): 198–213.

33 *making a plan:* D. Joel and I. Weiner, "The Organization of the Basal Ganglia–Thalamocortical Circuits: Open Interconnected Rather than Closed Segregated," *Neuroscience* 63, no. 2 (November 1994): 363–79.

34 *motor brain regions:* Redgrave, Prescott, and Gurney, "The Basal Ganglia"; Joel and Weiner, "The Organization of the Basal Ganglia–Thalamocortical Circuits."

36 *called the* substantia nigra: J. M. Fearnley and A. J. Lees, "Ageing and Parkinson's Disease: Substantia Nigra Regional Selectivity," *Brain* 114, no. 5 (October 1, 1991): 2283–301.

36 *move around a lot:* J. M. Delfs, L. Schreiber, and A. E. Kelley, "Microinjection of Cocaine into the Nucleus Accumbens Elicits Locomotor Activation in the Rat," *Journal of Neuroscience* 10, no. 1 (January 1, 1990): 303–10.

37 *well-worn movement patterns:* Fearnley and Lees, "Ageing and Parkinson's Disease"; P. Redgrave, M. Rodriguez, Y. Smith, M. C. Rodriguez-Oroz, S. Lehericy, H. Bergman, et al., "Goal-Directed and Habitual Control in the Basal Ganglia: Implications for Parkinson's Disease," *Nature Reviews Neuroscience* 11, no. 11 (November 2010): 760–72.

38 *more normally once again:* R. B. Godwin-Austen, C. C. Frears, E. B. Tomlinson, and H. W. L. Kok, "Effects of L-Dopa in Parkinson's Disease," *Lancet* 294, no. 7613 (July 26, 1969): 165–68.

38 *drug abuse, and binge eating:* V. Voon, P.-O. Fernagut, J. Wickens, C. Baunez, M. Rodriguez, N. Pavon, et al., "Chronic Dopaminergic Stimulation in Parkinson's Disease: From Dyskinesias to Impulse Control Disorders," *Lancet Neurology* 8, no. 12 (December 2009): 1140–49.

39   *cluster of unusual symptoms:* K. Barrett, "Treating Organic Abulia with Bromocriptine and Lisuride: Four Case Studies," *Journal of Neurology, Neurosurgery, and Psychiatry* 54, no. 8 (August 1, 1991): 718–21.

39   *motivations, emotions, and thoughts:* Barrett, "Treating Organic Abulia"; D. Laplane, M. Baulac, D. Widlöcher, and B. Dubois, "Pure Psychic Akinesia with Bilateral Lesions of Basal Ganglia," *Journal of Neurology, Neurosurgery, and Psychiatry* 47, no. 4 (1984): 377–85; S. E. Starkstein, M. L. Berthier, and R. Leiguarda, "Psychic Akinesia Following Bilateral Pallidal Lesions," *International Journal of Psychiatry in Medicine* 19, no. 2 (1989): 155–64; A. Lugaresi, P. Montagna, A. Morreale, and R. Gallassi, "'Psychic Akinesia' Following Carbon Monoxide Poisoning," *European Neurology* 30, no. 3 (1990): 167–69.

3. The Chemistry of Seduction

42   *2004 review paper:* R. A. Wise, "Dopamine, Learning, and Motivation," *Nature Reviews Neuroscience* 5, no. 6 (June 2004): 483–94.

43   *to recur also:* E. L. Thorndike, *The Elements of Psychology* (New York: A. G. Seiler, 1905), 394.

44   *eat there again:* I. L. Bernstein and M. M. Webster, "Learned Taste Aversions in Humans," *Physiology and Behavior* 25, no. 3 (September 1980): 363–66.

45   *once per second:* I. B. Witten, E. E. Steinberg, S. Y. Lee, T. J. Davidson, K. A. Zalocusky, M. Brodsky, et al., "Recombinase-Driver Rat Lines: Tools, Techniques, and Optogenetic Application to Dopamine-Mediated Reinforcement," *Neuron* 72, no. 5 (December 8, 2011): 721–33.

46   *again in the future:* J. N. Reynolds, B. I. Hyland, and J. R. Wickens, "A Cellular Mechanism of Reward-Related Learning," *Nature* 413, no. 6851 (September 6, 2001): 67–70.

46   *your "successful" behavior:* Wise, "Dopamine, Learning and Motivation."

46   *response to the bell alone:* I. P. Pavlov and G. V. Anrep, *Conditioned Reflexes* (Mineola, NY: Dover Publications, 2012), 448.

47   *humans back this up:* Ibid.

47   *experience of pleasure:* S. Peciña, K. S. Smith, and K. C. Berridge, "Hedonic Hot Spots in the Brain," *Neuroscientist: A Review Journal Bringing Neurobiology, Neurology, and Psychiatry* 12, no. 6 (December 2006): 500–11; K. C. Berridge, "'Liking' and 'Wanting' Food Rewards: Brain Substrates and Roles in Eating Disorders," *Physiology and Behavior* 97, no. 5 (July 14, 2009): 537–50.

48   *with the grape flavor:* A. Sclafani and J. W. Nissenbaum, "Robust Conditioned Flavor Preference Produced by Intragastric Starch Infusion in Rats," *American Journal of Physiology Regulatory, Integrative and Comparative Physiology* 255, no. 4 (October 1988): R672–R675.

48   *in the digestive tract:* G. Elizalde and A. Sclafani, "Starch-Based Conditioned Flavor Preferences in Rats: Influence of Taste, Calories, and CS-US Delay," *Appetite* 11, no. 3 (December): 179–200; A. Sclafani and K. Ackroff, "Glucose- and Fructose-Conditioned Flavor Preferences in Rats: Taste versus

Postingestive Conditioning," *Physiology and Behavior* 56, no. 2 (August 1994): 399–405.

48   *upper small intestine:* K. Ackroff, Y.-M. Yiin, A. Sclafani, "Post-Oral Infusion Sites That Support Glucose-Conditioned Flavor Preferences in Rats," *Physiology and Behavior* 99, no. 3 (March 3, 2010): 402–11.

48   *more dopamine spikes:* I. E. de Araujo, J. G. Ferreira, L. A. Tellez, X. Ren, and C. W. Yeckel, "The Gut–Brain Dopamine Axis: A Regulatory System for Caloric Intake," *Physiology and Behavior* 106, no. 3 (June 6, 2012): 394–99.

48n *they carry the signal:* A. Sclafani, K. Ackroff, and G. J. Schwartz, "Selective Effects of Vagal Deafferentation and Celiac-Superior Mesenteric Ganglionectomy on the Reinforcing and Satiating Action of Intestinal Nutrients," *Physiology and Behavior* 78, no. 2 (February 2003): 285–94.

49   *conditioned flavor preferences:* A. V. Azzara, R. J. Bodnar, A. R. Delamater, and A. Sclafani, "D1 but Not D2 Dopamine Receptor Antagonism Blocks the Acquisition of a Flavor Preference Conditioned by Intragastric Carbohydrate Infusions," *Pharmacology Biochemistry and Behavior* 68, no. 4 (April 2001): 709–20.

49   *using fat and protein:* C. Pérez, F. Lucas, and A. Sclafani, "Carbohydrate, Fat, and Protein Condition Similar Flavor Preferences in Rats Using an Oral-Delay Procedure," *Physiology and Behavior* 57, no. 3 (March 1995): 549–54.

49   *more reinforcing it is:* A. Ackroff and A. Sclatani, "Energy Density and Macronutrient Composition Determine Flavor Preference Conditioned by Intragastric Infusions of Mixed Diets," *Physiology and Behavior* 89, no. 2 (September 2006): 250–60.

49   *cooperate to reinforce behavior:* Sclafani and Ackroff, "Glucose- and Fructose-Conditioned Flavor."

50   *availability of calories:* C. Arnould and A. Ågmo, "The Importance of the Stomach for Conditioned Place Preference Produced by Drinking Sucrose in Rats," *Psychobiology* 27, no, 4 (December 1999): 541–46.

50   *administered into the stomach:* K. Ackroff and A. Sclafani, "Flavor Preferences Conditioned by Post-Oral Infusion of Monosodium Glutamate in Rats," *Physiology and Behavior* 104, no. 3 (September 1, 2011): 488–94.

50   *digestive distress are aversive:* Berridge, "'Liking' and 'Wanting' Food Rewards"; A. Sclafani, A. V. Azzara, K. Touzani, P. S. Grigson, and R. Norgren, "Parabrachial Nucleus Lesions Block Taste and Attenuate Flavor Preference and Aversion Conditioning in Rats," *Behavioral Neuroscience* 115, no. 4 (August 2001): 920–33; A. Dewan, R. Pacifico, R. Zhan, D. Rinberg, and T. Bozza, "Non-Redundant Coding of Aversive Odours in the Main Olfactory Pathway," *Nature* 498 (April 28, 2013): 486–89.

51   *seventy-five million years ago:* Berridge, "'Liking' and 'Wanting' Food Rewards."

51   *previously caused digestive distress:* Bernstein and Webster, "Learned Taste Aversions"; M. R. Yeomans, N. J. Gould, S. Mobini, J. Prescott, "Acquired Flavor Acceptance and Intake Facilitated by Monosodium Glutamate in Humans," *Physiology and Behavior* 94, nos. 4,5 (March 18, 2008): 958–66.

51   *our own species:* S. L. Johnson, L. McPhee, and L. L. Birch, "Conditioned Preferences: Young Children Prefer Flavors Associated with High Dietary Fat," *Physiology and Behavior* 50, no. 6 (December 1991): 1245–51; D. L. Kern,

L. McPhee, J. Fisher, S. Johnson, and L. L. Birch, "The Postingestive Consequences of Fat Condition Preferences for Flavors Associated with High Dietary Fat," *Physiology and Behavior* 54, no. 1 (July 1993): 71–76; D. A. Booth, P. Mather, and J. Fuller, "Starch Content of Ordinary Foods Associatively Conditions Human Appetite and Satiation, Indexed by Intake and Eating Pleasantness of Starch-Paired Flavours," *Appetite* 3, no. 2 (June 1982): 163–84.

51n   *like humans do:* R. R. Sakai, W. B. Fine, A. N. Epstein, and S. P. Frankmann, "Salt Appetite Is Enhanced by One Prior Episode of Sodium Depletion in the Rat," *Behavioral Neuroscience* 101, no. 5 (1987): 724–31.

52   *we'll return to later:* Lee, *The !Kung San;* A. M. Hurtado and K. Hill, *Ache Life History: The Ecology and Demography of a Foraging People* (New York: Aldine Transaction, 1996), 561.

53   *act on the same pathway:* S. Ferré, K. Fuxe, B. Fredholm, M. Morelli, and P. Popoli, "Adenosine–Dopamine Receptor–Receptor Interactions as an Integrative Mechanism in the Basal Ganglia," *Trends in Neurosciences* 20, no. 10 (October 1, 1997): 482–87.

54   *consequences, and withdrawal symptoms:* A. N. Gearhardt, W. R. Corbin, and K. D. Brownell, "Preliminary Validation of the Yale Food Addiction Scale," *Appetite* 52, no. 2 (April 2009): 430–36.

54   *criteria for food addiction:* Ibid.

54   *binge eating behavior:* A. Meule, "How Prevalent Is 'Food Addiction'?" *Frontiers in Psychiatry* 2 (November 2011): 61.

54   *sheds light on the question:* Gearhardt, Corbin, and Brownell, "Yale Food Addiction Scale."

56   *like its cousin caffeine:* H. J. Smit and R. J. Blackburn, "Reinforcing Effects of Caffeine and Theobromine as Found in Chocolate," *Psychopharmacology* (Berlin) 181, no. 1 (August 1, 2005): 101–106.

56   *craved food among women:* A. J. Hill and L. Heaton-Brown, "The Experience of Food Craving: A Prospective Investigation in Healthy Women," *Journal of Psychosomatic Research* 38, no. 8 (November 1994): 801–14.

58   *foods they like:* M. R. Yeomans, "Palatability and the Micro-structure of Feeding in Humans: The Appetizer Effect," *Appetite* 27, no. 2 (October 1996): 119–33; C. de Graaf, L. S. de Jong, and A. C. Lambers, "Palatability Affects Satiation but Not Satiety," *Physiology and Behavior* 66, no. 4 (June 1999): 681–88; M. O. Monneuse, F. Bellisle, J. Louis-Sylvestre, "Responses to an Intense Sweetener in Humans: Immediate Preferences and Delayed Effects on Intake," *Physiology and Behavior* 49, no. 2 (February 1991): 325–30; E. M. Bobroff and H. R. Kissileff, "Effects of Changes in Palatability on Food Intake and the Cumulative Food Intake Curve of Man," *Appetite* 7, no. 1 (March 1986): 85–96.

58   *meals they describe as bland:* J. M. de Castro, "Eating Behavior: Lessons from the Real World of Humans," *Nutrition* (Los Angeles) 16, no. 10 (October 2000): 800–13.

58   *unintentionally addressed this question:* S. A. Hashim, T. B. van Itallie, "Studies in

Normal and Obese Subjects with a Monitored Food Dispensing Device," *Annals of the New York Academy of Sciences* 131, no. 1 (1965): 654–61.

60   *lose excess fat:* Ibid.; R. G. Campbell, S. A. Hashim, and T. B. van Itallie, "Studies of Food-Intake Regulation in Man," *New England Journal of Medicine* 285, no. 25 (1971): 1402–407; M. Cabanac and E. F. Rabe, "Influence of a Monotonous Food on Body Weight Regulation in Humans," *Physiology and Behavior* 17, no. 4 (October 1976): 675–78.

61   *months at a time:* S. K. Kon and A. Klein, "The Value of Whole Potato in Human Nutrition," *Biochemical Journal* 22, no. 1 (1928): 258–60; M. S. Rose and L. F. Cooper, "The Biological Efficiency of Potato Nitrogen," *Journal of Biological Chemistry* 30 (1917): 201–204.

61   *maintain his weight:* C. Voigt, "20 Potatoes a Day," http://www.20potatoesaday.com/.

61   *rapid weight loss:* "EAT MOAR TATERS!" 2012, http://www.marksdailyapple.com/forum/thread67137.html.

62   *ancient relatives the jellyfish:* M. C. Johnson and K. L. Wuensch, "An Investigation of Habituation in the Jellyfish *Aurelia aurita*," *Behavioral and Neural Biology* 61, no. 1 (1994): 54–59.

62   *experiments in human infants:* H. S. Bashinski, J. S. Werner, and J. W. Rudy, "Determinants of Infant Visual Fixation: Evidence for a Two-Process Theory," *Journal of Experimental Child Psychology* 39, no. 3 (June 1985): 580–98.

62   *foods for lunch:* B. J. Rolls, E. T. Rolls, E. A. Rowe, and K. Sweeney, "Sensory Specific Satiety in Man," *Physiology and Behavior* 27, no. 1 (July 1981): 137–42.

63   *large variety of foods:* B. J. Rolls, P. M. van Duijvenvoorde, and E. A. Rowe, "Variety in the Diet Enhances Intake in a Meal and Contributes to the Development of Obesity in the Rat," *Physiology and Behavior* 31, no. 1 (July 1983): 21–27; H. A. Raynor and L. H. Epstein, "Dietary Variety, Energy Regulation, and Obesity," *Psychological Bulletin* 127, no. 3 (May 2001): 325–41; R. J. Stubbs, A. M. Johnstone, N. Mazlan, S. E. Mbaiwa, and S. Ferris, "Effect of Altering the Variety of Sensorially Distinct Foods, of the Same Macronutrient Content, on Food Intake and Body Weight in Men," *European Journal of Clinical Nutrition* 55, no. 1 (January 2001): 19–28; B. J. Rolls, E. A. Rowe, E. T. Rolls, B. Kingston, A. Megson, and R. Gunary, "Variety in a Meal Enhances Food Intake in Man," *Physiology and Behavior* 26, no. 2 (February 1981): 215–21.

64   *"the munchies":* R. W. Foltin, M. W. Fischman, and M. F. Byrne, "Effects of Smoked Marijuana on Food Intake and Body Weight of Humans Living in a Residential Laboratory," *Appetite* 11, no. 1 (August 1988): 1–14.

64   *regulates food reward:* G. Jager and R. F. Witkamp, "The Endocannabinoid System and Appetite: Relevance for Food Reward," *Nutrition Research Reviews* (June 16, 2014): 1–14.

64   *marijuana increases food intake:* R. W. Foltin, J. V. Brady, and M. W. Fischman, "Behavioral Analysis of Marijuana Effects on Food Intake in Humans," *Pharmacology Biochemistry and Behavior* 25, no. 3 (September 1986): 577–82; E. L. Abel, "Effects of Marihuana on the Solution of Anagrams, Memory and Appetite," *Nature* 231, no. 5100 (May 1971): 260–61; E. G. Williams and C. K.

Himmelsbach, "Studies on Marihuana and Pyrahexyl Compound," *Public Health Reports* 61 (July 19, 1946): 1059–83.

64   *negative side effects:* R. Christensen, P. K. Kristensen, E. M. Bartels, H. Bliddal, and A. Astrup, "Efficacy and Safety of the Weight-Loss Drug Rimonabant: A Meta-Analysis of Randomised Trials," *Lancet* 370, no. 9600 (November 17, 2007): 1706–13; L. van Gaal, X. Pi-Sunyer, J.-P. Després, C. McCarthy, and A. Scheen, "Efficacy and Safety of Rimonabant for Improvement of Multiple Cardiometabolic Risk Factors in Overweight/Obese Patients: Pooled 1-Year Data from the Rimonabant in Obesity (RIO) Program," *Diabetes Care* 31, suppl. 2 (February 2008): S229–40; L. F. van Gaal, A. M. Rissanen, A. J. Scheen, O. Ziegler, and S. Rössner, "Effects of the Cannabinoid-1 Receptor Blocker Rimonabant on Weight Reduction and Cardiovascular Risk Factors in Overweight Patients: 1-Year Experience from the RIO-Europe Study," *Lancet* 365, no. 9468 (April 22, 2005): 1389–97.

64   *anxiety, and suicidal thoughts:* Van Gaal, Pi-Sunyer, Després, McCarthy, and Scheen, "Efficacy and Safety of Rimonabant."

66   *people who are lean:* B. E. Saelens and L. H. Epstein, "Reinforcing Value of Food in Obese and Non-Obese Women," *Appetite* 27, no. 1 (August 1996): 41–50.

66   *hunger is the same:* J. L. Temple, C. M. Legierski, A. M. Giacomelli, S.-J. Salvy, and L. H. Epstein, "Overweight Children Find Food More Reinforcing and Consume More Energy Than Do Nonoverweight Children," *American Journal of Clinical Nutrition* 87, no. 5 (May 1, 2008): 1121–27.

66   *lab and at home:* Ibid.; L. H. Epstein, K. A. Carr, H. Lin, K. D. Fletcher, and J. N. Roemmich, "Usual Energy Intake Mediates the Relationship Between Food Reinforcement and BMI," *Obesity* 20, no. 9 (September 1, 2012): 1815–19; L. H. Epstein, J. L. Temple, B. J. Neaderhiser, R. J. Salis, R. W. Erbe, and J. J. Leddy, "Food Reinforcement, the Dopamine D2 Receptor Genotype, and Energy Intake in Obese and Nonobese Humans," *Behavioral Neuroscience* 121, no. 5 (October 2007): 877–86.

66   *age group they examined:* C. Hill, J. Saxton, L. Webber, J. Blundell, and J. Wardle, "The Relative Reinforcing Value of Food Predicts Weight Gain in a Longitudinal Study of 7–10-Y-Old Children," *American Journal of Clinical Nutrition* 90, no. 2 (August 1, 2009): 276–81; K. A. Carr, H. Lin, K. D. Fletcher, and L. H. Epstein, "Food Reinforcement, Dietary Disinhibition and Weight Gain in Nonobese Adults," *Obesity* 22, no. 1 (January 1, 2014): 254–59.

66   *gained only half a pound:* Carr, Lin, Fletcher, and Epstein, "Food Reinforcement, Dietary Disinhibition."

68   *highly susceptible to overeating:* B. Y. Rollins, K. K. Dearing, and L. H. Epstein, "Delay Discounting Moderates the Effect of Food Reinforcement on Energy Intake among Non-Obese Women," *Appetite* 55, no. 3 (December 2010): 420–25; B. M. Appelhans, K. Woolf, S. L. Pagoto, K. L. Schneider, M. C. Whited, and R. Liebman, "Inhibiting Food Reward: Delay Discounting, Food Reward Sensitivity, and Palatable Food Intake in Overweight and Obese Women," *Obesity* (Silver Spring, MD) 19, no. 11 (November 2011): 2175–82.

68   *susceptible to weight gain:* Carr, Lin, Fletcher, and Epstein, "Food Reinforcement, Dietary Disinhibition"; C. Nederkoorn, K. Houben, W. Hofmann, A. Roefs,

and A. Jansen, "Control Yourself or Just Eat What You Like? Weight Gain over a Year Is Predicted by an Interactive Effect of Response Inhibition and Implicit Preference for Snack Foods," *Health Psychology: Official Journal of the Division of Health Psychology of the American Psychological Association* 29, no. 4 (July 2010): 389–93.

4. The United States of Food Reward

70 *heart disease, diabetes, and obesity:* Lieberman, *The Story of the Human Body*; Lindeberg, *Food and Western Disease.*

70 *preparation, and eating practices:* Lee, *The !Kung San.*

70 *from only fourteen species:* Ibid.

70 *liver being particularly prized:* Ibid.

70 *roasted cashews or almonds:* Ibid.

71 *Lee describes as follows:* Ibid.

72 *foods in a particular location:* Ibid.

72 *"without much enthusiasm":* Ibid.

72 *reproductive years and declined thereafter:* Ibid.

72 *beginning in 1964:* N. A. Chagnon and E. O. Wilson, *Yanomamö: The Last Days of Eden: The Celebrated Anthropologist's Pioneer Work among a Now-Imperiled Amazon Tribe* (San Diego: Jovanovich, 1992).

72 *"animal food alike":* Ibid.

73 *remarkably low throughout life:* J. J. Mancilha-Carvalho, R. de Oliveira, R. J. Esposito, "Blood Pressure and Electrolyte Excretion in the Yanomamo Indians, an Isolated Population," *Journal of Human Hypertension* 3, no. 5 (October 1989): 309–14.

73 *adequate food supplies:* Ibid.

73 *season and location:* Lee, *The !Kung San.*

74 *cooking techniques difficult:* S. Williams, *Food in the United States, 1820s–1890*, 1st ed. (Santa Barbara, CA: Greenwood, 2006), 264; M. J. Elias, *Food in the United States, 1890–1945* (Santa Barbara, CA: Greenwood Press/ABC-CLIO, 2009).

74 *stoves in the 1920s:* Ibid; Chagnon and Wilson, *Yanomamö.*

75 *fifteen thousand items in 1980:* Food Marketing Institute, "Supermarket Facts," 2013, http://www.fmi.org/research-resources/supermarket-facts.

75 *difficult to overstate:* USDA, "Major Trends in U.S. Food Supply, 1909–99," *FoodReview* 21, no. 1 (2000): 8–15; USDA, "USDA Economic Research Service— Food Availability."

75 *other half eating out:* USDA, "USDA Economic Research Service—Food Availability."

77 *cost of sweetened beverages:* Elias, *Food in the United States.*

77 *cold, refreshing soda:* Ibid.

77 *from 1822 to 2005:* USDA, "USDA Economic Research Service—Food Availability"; *Statistical Abstract of the United States* (US Government Printing Office, 1907), 763; *Statistical Abstract of the United States* (US Government Printing Office, 1920), 892.

80 *two thousand years ago:* R. I. Curtis, "Umami and the Foods of Classical

Antiquity," *American Journal of Clinical Nutrition* 90, no. 3 (September 2009): 712S–718S.

80   *concentrated forms of glutamate:* K. Kurihara, "Glutamate: From Discovery as a Food Flavor to Role as a Basic Taste (Umami)," *American Journal of Clinical Nutrition* 90, no. 3 (September 2009): 719S–722S.

80   *dozens of ingredients:* Schlosser, *Fast Food Nation*; D. A. Kessler, *The End of Overeating: Taking Control of the Insatiable American Appetite* (Emmaus, PA: Rodale, 2009).

80   *can contribute to overeating:* P. J. Rogers, N. J. Richardson, and N. A. Elliman, "Overnight Caffeine Abstinence and Negative Reinforcement of Preference for Caffeine-Containing Drinks," *Psychopharmacology* (Berlin) 120, no. 4 (August 1, 1995): 457–62; C. L. Cunningham and J. S. Niehus, "Flavor Preference Conditioning by Oral Self-Administration of Ethanol," *Psychopharmacology* (Berlin) 134, no. 3 (December 1, 1997): 293–302.

80n   *failed to support this idea:* R. S. Geha, A. Beiser, C. Ren, R. Patterson, P. A. Greenberger, L. C. Grammer, et al., "Review of Alleged Reaction to Monosodium Glutamate and Outcome of a Multicenter Double-Blind Placebo-Controlled Study," *Journal of Nutrition* 130, no. 4 (April 1, 2000): 1058S–1062S.

80n   *brain levels, of glutamate significantly:* R. Walker and J. R. Lupien, "The Safety Evaluation of Monosodium Glutamate," *Journal of Nutrition* 130, 4S suppl. (April 2000): 1049S–52S.

81   *a repeat customer:* E. Schlosser, *Fast Food Nation: The Dark Side of the All-American Meal* (Boston: Mariner Books / Houghton Mifflin Harcourt, 2012); M. Moss, *Salt, Sugar, Fat: How the Food Giants Hooked Us* (New York: Random House, 2013).

82   *version of natural eggs:* O. Koehler and A. Zagarus, "Beiträge zum Brutverhalten des Halsbandregepfeifers (Charadrius h. hiaticula L.)," *Beitr Zur Fortpflanzungs-biologie Vögel* 13 (1937): 1–9.

82   *their own bodies:* N. Tinbergen, *The Study of Instinct* (Oxford, UK: Oxford University Press, 1991), 256.

82   *"the natural situation":* Ibid.

83   *a normal chick:* W. Wickler, *Mimicry in Plants and Animals* (New York: McGraw-Hill, 1968), 260.

83   *contributes to our diet:* US Department of Agriculture and US Department of Health and Human Services, *Dietary Guidelines for Americans, 2010* (US Government Printing Office, 2010).

84   *salt in the US diet:* Centers for Disease Control and Prevention, "Vital Signs: Food Categories Contributing the Most to Sodium Consumption—United States, 2007–2008," cited June 24, 2014, http://www.cdc.gov/mmwr/preview/mmwrhtml/mm6105a3.htm?s_cid=mm6105a3_w.

85   *$4 billion on top of that:* D. Bailin, G. Goldman, and P. Phartiyal, *Sugar-Coating Science: How the Food Industry Misleads Consumers on Sugar* (Cambridge, MA: Union of Concerned Scientists, 2014).

85   *$1 billion in 2012:* Nationals Institutes of Health, "NIH Categorical Spending—

NIH Research Portfolio Online Reporting Tools (RePORT)," cited March 11, 2016, https://report.nih.gov/categorical_spending.aspx.

85   *request the advertised products:* National Research Council, *Food Marketing to Children and Youth.*

85   *exposures per year:* C. R. Dembek, J. L. Harris, and M. B. Schwartz, *Where Children and Adolescents View Food and Beverage Ads on TV: Exposure by Channel and Program* (New Haven, CT: Yale Rudd Center for Food Policy and Obesity, 2013).

85   *4,300 per year:* National Research Council, *Food Marketing to Children and Youth: Threat or Opportunity?* (Washington, D.C.: National Academies Press, 2006).

85   *most often in advertisements:* Bailin, Goldman, and Phartiyal, *Sugar-Coating Science;* Dembek, Harris, and Schwartz, "Where Children and Adolescents View Food"; National Research Council, *Food Marketing to Children and Youth.*

5. The Economics of Eating

87   *2.6 million years ago:* R. G. Klein, *The Human Career: Human Biological and Cultural Origins,* 3rd ed. (Chicago: University of Chicago Press, 2009), 1024.

88n   *reaching ninety-five pounds:* F. Marlowe, *The Hadza: Hunter-Gatherers of Tanzania* (Berkeley: University of California Press, 2010), 336.

90   *Brian Wood, and others:* Ibid.

92   *human hunter-gatherers:* Ibid.; B. Winterhalder and E. A. Smith, "Analyzing Adaptive Strategies: Human Behavioral Ecology at Twenty-Five," *Evolutionary Anthropology: Issues, News, and Reviews* 9, no. 2 (January 1, 2000): 51–72; K. Hill, H. Kaplan, K. Hawkes, and A. M. Hurtado, "Foraging Decisions Among Aché Hunter-Gatherers: New Data and Implications for Optimal Foraging Models," *Ethology and Social Biology* 8, no. 1 (1987): 1–36; E. A. Smith, R. L. Bettinger, C. A. Bishop, V. Blundell, E. Cashdan, M. J. Casimir, et al., "Anthropological Applications of Optimal Foraging Theory: A Critical Review," *Current Anthropology* 24, no. 5 (December 1, 1983): 625–51.

92   *obtain and process it:* Winterhalder and Smith, "Analyzing Adaptive Strategies"; Hill, Kaplan, Hawkes, and Hurtado, "Foraging Decisions"; Smith, Bettinger, Bishop, Blundell, Cashdan, Casimir, et al., "Anthropological Applications."

92   *behavior of human hunter-gatherers:* Hurtado and Hill, *Ache Life History;* Marlowe, *The Hadza;* Winterhalder and Smith "Analyzing Adaptive Strategies"; Hill, Kaplan, Hawkes, and Hurtado, "Foraging Decisions"; K. Hawkes, K. Hill, and J. F. O'Connell, "Why Hunters Gather: Optimal Foraging and the Aché of Eastern Paraguay," *American Ethnologist* 9, no. 2 (May 1, 1982): 379–98; B. M. Wood and F. W. Marlowe, "Toward a Reality-Based Understanding of Hadza Men's Work: A Response to Hawkes et al.," *Human Nature* (Hawthorne, NY) 25, no. 4 (December 2014): 620–30.

92   *low in calories:* Lee, *The !Kung San;* Marlowe, *The Hadza;* Hawkes, Hill, and O'Connell, "Why Hunters Gather."

92n   *justify a move:* Winterhalder and Smith, "Analyzing Adaptive Strategies"; Hill,

Kaplan, Hawkes, and Hurtado, "Foraging Decisions"; Smith, Bettinger, Bishop, Blundell, Cashdan, Casimir, et al., "Anthropological Applications."

93 *locate a good nest:* F. W. Marlowe, J. C. Berbesque, B. Wood, A. Crittenden, C. Porter, and A. Mabulla, "Honey, Hadza, Hunter-Gatherers, and Human Evolution," *Journal of Human Evolution* 71 (2014): 119–28.

93 *can impact its value:* Marlowe and Berbesque, "Tubers as Fallback Foods"; Hill, Kaplan, Hawkes, and Hurtado, "Foraging Decisions."

94 *among Aché hunter-gatherers:* Hill, Kaplan, Hawkes, and Hurtado, "Foraging Decisions."

94 *risk of falling:* Ibid.

95 *fatter with age:* Marlowe, *The Hadza.*

95 *they are rarely obese:* R. G. Bribiescas, "Serum Leptin Levels and Anthropometric Correlates in Ache Amerindians of Eastern Paraguay," *American Journal of Physical Anthropology* 115, no. 4 (August 2001): 297–303.

95n *taken into account:* H. Pontzer, D. A. Raichlen, B. M. Wood, A. Z. P. Mabulla, S. B. Racette, and F. W. Marlowe, "Hunter-Gatherer Energetics and Human Obesity," *PLOS ONE* 7, no. 7 (2012): e40503.

96 *wants more energy:* Hurtado and Hill, *Ache Life History;* K. Hawkes, J. F. O'Connell, K. Hill, and E. L. Charnov, "How Much Is Enough? Hunters and Limited Needs," *Ethology and Social Biology* 6, no. 1 (1985): 3–15.

96n *part of the diet:* L. Cordain, J. B. Miller, S. B. Eaton, N. Mann, S. H. Holt, and J. D. Speth, "Plant-Animal Subsistence Ratios and Macronutrient Energy Estimations in Worldwide Hunter-Gatherer Diets," *American Journal of Clinical Nutrition* 71, no. 3 (March 2000): 682–92.

97 *have drastically declined:* L. S. Lieberman, "Evolutionary and Anthropological Perspectives on Optimal Foraging in Obesogenic Environments," *Appetite* 47, no. 1 (July 2006): 3–9.

97n *hunter-gatherer and agricultural populations:* M. N. Cohen, *Paleopathology at the Origins of Agriculture* (Orlando, FL: Academic Press, 1984), 644.

98 *all times of day:* Rising, Alger, Boyce, Seagle, Ferraro, Fontvieille, et al., "Food Intake"; Larson, Tataranni, Ferraro, and Ravussin, "Ad Libitum Food Intake"; Larson, Rising, Ferraro, and Ravussin, "Spontaneous Overfeeding."

98 *cabinet six feet away:* B. Wansink, "Environmental Factors That Increase the Food Intake and Consumption Volume of Unknowing Consumers," *Annual Review of Nutrition* 24 (2004): 455–79; B. Wansink, *Mindless Eating: Why We Eat More Than We Think* (New York: Bantam, 2010).

98 *"locate and overeat mangos":* Wansink, *Mindless Eating.*

99 *income to 10 percent:* USDA, *USDA Economic Research Service—Food Availability.*

99 *expand in size since then:* M. J. Elias, *Food in the United States, 1890–1945* (Santa Barbara, CA: Greenwood Press / ABC-CLIO, 2009).

100 *convenience meal: fast food:* USDA, *USDA Economic Research Service—Food Availability.*

100 *maximum consumer appeal:* Moss, *Salt, Sugar, Fat.*

101 *cost-benefit decisions:* C. Padoa-Schioppa and J. A. Assad, "Neurons in Orbito-frontal Cortex Encode Economic Value," *Nature* 441, no. 7090 (May 11, 2006): 223–26.

103 *how the brain computes it:* P. Glimcher, "Understanding the Hows and Whys of Decision-Making: From Expected Utility to Divisive Normalization," Cold Spring Harbor Symposia on Quantitative Biology, January 30, 2015; C. Padoa-Schioppa, "Neurobiology of Economic Choice: A Good-Based Model," *Annual Review of Neuroscience* 34 (2011): 333–59.

103 *value of specific options:* Padoa-Schioppa and Assad, "Neurons in Orbitofrontal Cortex."

103 *costs required to obtain it:* Padoa-Schioppa, "Neurobiology of Economic Choice"; A. P. Raghuraman and C. Padoa-Schioppa, "Integration of Multiple Determinants in the Neuronal Computation of Economic Values," *Journal of Neuroscience* 34, no. 35 (August 27, 2014): 11583–603; S. W. Kennerley, A. F. Dahmubed, A. H. Lara, and J. D. Wallis, "Neurons in the Frontal Lobe Encode the Value of Multiple Decision Variables," *Journal of Cognitive Neuroscience* 21, no. 6 (June 2009): 1162–78.

103 *ventromedial prefrontal cortex:* Padoa-Schioppa, "Neurobiology of Economic Choice"; T. A. Hare, W. Schultz, C. F. Camerer, J. P. O'Doherty, and A. Rangel, "Transformation of Stimulus Value Signals into Motor Commands During Simple Choice," *Proceedings of the National Academy of Sciences of the United States of America* 108, no. 44 (November 1, 2011): 18120–25; A. Rangel, "Regulation of Dietary Choice by the Decision-Making Circuitry," *Nature Neuroscience* 16, no. 12 (December 2013): 1717–24.

105 *follow up on the decision:* Hare, Schultz, Camerer, O'Doherty, and Rangel, "Transformation of Stimulus"; X. Cai and C. Padoa-Schioppa, "Contributions of Orbitofrontal and Lateral Prefrontal Cortices to Economic Choice and the Good-to-Action Transformation," *Neuron* 81, no. 5 (March 5, 2014): 1140–51.

106 *compute value on the fly:* E. A. West, J. T. DesJardin, K. Gale, and L. Malkova, "Transient Inactivation of Orbitofrontal Cortex Blocks Reinforcer Devaluation in Macaques," *Journal of Neuroscience* 31, no. 42 (October 19, 2011): 15128–35; M. Gallagher, R. W. McMahan, and G. Schoenbaum, "Orbitofrontal Cortex and Representation of Incentive Value in Associative Learning," *Journal of Neuroscience* 19, no. 15 (August 1, 1999): 6610–4; A. Tsuchida, B. B. Doll, and L. K. Fellows, "Beyond Reversal: A Critical Role for Human Orbitofrontal Cortex in Flexible Learning from Probabilistic Feedback," *Journal of Neuroscience: The Official Journal of the Society for Neuroscience* 30, no. 50 (December 15, 2010): 16868–75.

106 *results from OFC damage:* Clarke, Robbins, and Roberts, "Lesions of the Medial Striatum"; J. S. Snowden, D. Neary, and D. M. A. Mann, "Frontotemporal Dementia," *British Journal of Psychiatry* 180 (February 2002): 140–43.

106 *overeating and weight gain:* J. D. Woolley, M.-L. Gorno-Tempini, W. W. Seeley, K. Rankin, S. S. Lee, B. R. Matthews, et al., "Binge Eating Is Associated with Right Orbitofrontal-Insular-Striatal Atrophy in Frontotemporal Dementia," *Neurology* 69, no. 14 (October 2, 2007): 1424–33.

106 *doesn't reach the OFC:* Ibid.

106n *can cause similar deficits:* H. F. Clarke, T. W. Robbins, and A. C. Roberts, "Lesions of the Medial Striatum in Monkeys Produce Perseverative Impairments During

Reversal Learning Similar to Those Produced by Lesions of the Orbitofrontal Cortex," *Journal of Neuroscience* 28, no. 43 (October 22, 2008): 10972–82.

107   *dopamine is chemically replaced:* Q. Y. Zhou and R. D. Palmiter, "Dopamine-Deficient Mice Are Severely Hypoactive, Adipsic, and Aphagic," *Cell* 83, no. 7 (December 29, 1995): 1197–209.

107   *work for a reward:* J. D. Salamone and M. Correa, "The Mysterious Motivational Functions of Mesolimbic Dopamine," *Neuron* 76, no. 3 (November 8, 2012): 470–85.

107   *reward is small or uncertain:* M. C. Wardle, M. T. Treadway, L. M. Mayo, D. H. Zald, and H. de Wit, "Amping Up Effort: Effects of d-Amphetamine on Human Effort-Based Decision-Making," *Journal of Neuroscience* 31, no. 46 (November 16, 2011): 16597–602.

108   *marshmallows in fifteen minutes:* W. Mischel and E. B. Ebbesen, "Attention in Delay of Gratification," *Journal of Personality and Social Psychology* 16, no. 2 (1970): 329–37.

109   *shortchanging their future selves:* Mischel and Ebbesen, "Attention in Delay of Gratification"; W. Mischel, E. B. Ebbesen, and A. Raskoff Zeiss, "Cognitive and Attentional Mechanisms in Delay of Gratification," *Journal of Personality and Social Psychology* 21, no. 2 (February 1972): 204–18.

109   *slimmer thirty years later:* T. R. Schlam, N. L. Wilson, Y. Shoda, W. Mischel, and O. Ayduk, "Preschoolers' Delay of Gratification Predicts Their Body Mass 30 Years Later," *Journal of Pediatrics* 162, no. 1 (January 2013): 90–93.

109   *likely to be obese:* D. P. Jarmolowicz, J. B. C. Cherry, D. D. Reed, J. M. Bruce, J. M. Crespi, J. L. Lusk, et al., "Robust Relation Between Temporal Discounting Rates and Body Mass," *Appetite* 78 (July 2014): 63–67; L. H. Epstein, N. Jankowiak, K. D. Fletcher, K. A. Carr, C. Nederkoorn, H. Raynor, et al., "Women Who Are Motivated to Eat and Discount the Future Are More Obese," *Obesity* (Silver Spring, MD) 22, no. 6 (June 2014): 1394–99.

109   *credit card debt:* Audrain-McGovern, Rodriguez, Epstein, Cuevas, Rodgers, and Wileyto, "Does Delay Discounting"; N. M. Petry, "Delay Discounting of Money and Alcohol in Actively Using Alcoholics, Currently Abstinent Alcoholics, and Controls," *Psychopharmacology* (Berlin) 153, no. 3 (March 1, 2001): 243–50; S. F. Coffey, G. D. Gudleski, M. E. Saladin, and K. T. Brady, "Impulsivity and Rapid Discounting of Delayed Hypothetical Rewards in Cocaine-Dependent Individuals," *Experimental and Clinical Psychopharmacology* 11, no. 1 (2003): 18–25; J. MacKillop, M. T. Amlung, L. R. Few, L. A. Ray, L. H. Sweet, and M. R. Munafò, "Delayed Reward Discounting and Addictive Behavior: A Meta-Analysis," *Psychopharmacology* (Berlin) 216, no. 3 (March 4, 2011): 305–21; S. M. Alessi and N. M. Petry, "Pathological Gambling Severity Is Associated with Impulsivity in a Delay Discounting Procedure," *Behavioural Processes* 64, no. 3 (October 31, 2003): 345–54; S. Meier and C. Sprenger, "Present-Biased Preferences and Credit Card Borrowing," *American Economic Journal: Applied Economics* 2, no. 1 (January 1, 2010): 193–210.

110   *more than our future selves:* Hurtado and Hill, *Ache Life History.*

110n   *contribute to addiction risk:* J. Audrain-McGovern, D. Rodriguez, L. H. Epstein, J. Cuevas, K. Rodgers, and E. P. Wileyto, "Does Delay Discounting Play an

Etiological Role in Smoking or Is It a Consequence of Smoking?," *Drug and Alcohol Dependence* 103, no. 3 (August 1, 2009): 99–106.

111   *episodic future thinking:* C. M. Atance and D. K. O'Neill, "Episodic Future Thinking," *Trends in Cognitive Sciences* 5, no. 12 (December 1, 2001): 533–39.

111   *decision-making process:* J. Peters and C. Büchel, "Episodic Future Thinking Reduces Reward Delay Discounting Through an Enhancement of Prefrontal-Mediotemporal Interactions," *Neuron* 66, no. 1 (April 15, 2010): 138–48.

111   *overweight children as well:* T. O. Daniel, C. M. Stanton, and L. H. Epstein, "The Future Is Now: Reducing Impulsivity and Energy Intake Using Episodic Future Thinking," *Psychological Science* 24, no. 11 (November 1, 2013): 2339–42; T. O. Daniel, M. Said, C. M. Stanton, and L. H. Epstein, "Episodic Future Thinking Reduces Delay Discounting and Energy Intake in Children," *Eating Behaviors* 18 (August 2015): 20–24.

6. The Satiety Factor

113   *"uncommonly large fat deposits":* B. Mohr, "Hypertrophie der Hypophysis cerebri und dadurch bedingter Druck auf die Hirngrundflashe, insbesondere auf die Sehverven, das Chiasma derselben und den linkseitigen Hirnschenkel," *Wochen-schrift Ges Heilkunde* 6 (1840): 565–71; B. Mohr, "Neuropathology Communication from Dr. Mohr, Privat Docent in Würzburg. 1840," *Obesity Research* 1, no. 4 (July 1993): 334–35.

114   *location Mohr had described:* A. Frohlich, "Dr. Alfred Frohlich stellt einen fall von tumor der hypophyse ohne akromegalie vor," *Wien Klin Rundsch* 15 (1902): 883–86, 906–908.

114   *known as Fröhlich's syndrome:* B. M. King, "The Rise, Fall, and Resurrection of the Ventromedial Hypothalamus in the Regulation of Feeding Behavior and Body Weight," *Physiology and Behavior* 87, no. 2 (February 28, 2006): 221–44.

115   *Fröhlich's study was published:* Ibid.

115   *damage to the pituitary itself:* J. Erdheim, "Uber Hypophysenganggeschwulste und Hirncholestcatome," *Akad Wiss Wien* 113 (1904): 537–726.

115   *obesity in dogs and rats:* P. Bailey and F. Bremer, "Experimental Diabetes Insipidus," *Archives of Internal Medicine* 28, no. 6 (December 1, 1921): 773–803; P. Smith, "The Disabilities Caused by Hypophysectomy and Their Repair: The Tuberal (Hypothalamic) Syndrome in the Rat," *Journal of the American Medical Association* 88, no. 3 (January 15, 1927): 158–61; P. E. Smith, "Hypophysectomy and a Replacement Therapy in the Rat," *American Journal of Anatomy* 45, no. 2 (March 1, 1930): 205–73; A. W. Hetherington, "Obesity in the Rat Following the Injection of Chromic Acid into the Hypophysis," *Endocrinology* 26, no. 2 (February 1, 1940): 264–68.

115   *ventromedial hypothalamic nucleus (VMN):* A. Hetherington, "The Relation of Various Hypothalamic Lesions to Adiposity and Other Phenomena in the Rat," *American Journal of Physiology* 133 (1941): 326–27; A. W. Hetherington, "The Production of Hypothalamic Obesity in Rats Already Displaying Chronic Hypopituitarism," *American Journal of Physiology* 140, no. 1 (1943): 89–92.

115 *in the early 1940s:* J. R. Brobeck, J. Tepperman, and C. N. H. Long, "Experimental Hypothalamic Hyperphagia in the Albino Rat," *Yale Journal of Biology and Medicine* 15, no. 6 (July 1943): 831–53.

117 *excessive food intake:* Ibid.

117n *rats to become obese:* King, "The Rise, Fall, and Resurrection."

118 *happened to be a male:* E. R. Shell, *The Hungry Gene: The Inside Story of the Obesity Industry,* 1st trade paper ed. (New York: Grove Press, 2003), 304.

118 *spontaneous genetic mutation:* A. M. Ingalls, M. M. Dickie, and G. D. Snell, "*Obese,* a New Mutation in the House Mouse," *Journal of Heredity* 41, no. 12 (December 1950): 317–18.

118 *reminiscent of human obesity:* G. A. Bray and D. A. York, "Hypothalamic and Genetic Obesity in Experimental Animals: An Autonomic and Endocrine Hypothesis," *Physiological Reviews* 59, no. 3 (July 1, 1979): 719–809.

118 *the* obese *mouse:* L. M. Zucker and T. F. Zucker, "*Fatty,* a New Mutation in the Rat," *Journal of Heredity* 52, no. 6 (1961): 275–78.

118 *their prodigious appetites:* B. Ga, "The *Zucker-Fatty* Rat: A Review," *Federation Proceedings* 36, no. 2 (February 1977): 148–53.

119 *leaving its twin intact:* G. R. Hervey, "The Effects of Lesions in the Hypothalamus in Parabiotic Rats," *Journal of Physiology* 145, no. 2 (March 3, 1959): 336–52.

120 *developed by Gordon Kennedy:* G. C. Kennedy, "The Role of Depot Fat in the Hypothalamic Control of Food Intake in the Rat," *Proceedings of the Royal Society of London B: Biological Sciences* 140, no. 901 (January 15, 1953): 578–92.

120 *constrain appetite and adiposity:* Hervey, "The Effects of Lesions."

121 *weight remained stable:* D. L. Coleman, "Effects of Parabiosis of Obese with Diabetes and Normal Mice," *Diabetologia* 9, no. 4 (August 1973): 294–98.

122 *other twin lost fat:* R. B. Harris and R. J. Martin, "Specific Depletion of Body Fat in Parabiotic Partners of Tube-Fed Obese Rats," *American Journal of Physiology* 247, no. 2 pt. 2 (August 1984): R380–R386.

122 *non-overfed twin:* R. B. Harris and R. J. Martin, "Site of Action of Putative Lipostatic Factor: Food Intake and Peripheral Pentose Shunt Activity," *American Journal of Physiology* 259, no. 1 pt. 2 (July 1990): R45–R52.

122 *known at the time:* Harris and Martin, "Specific Depletion of Body Fat."

122n *members of control pairs:* R. B. S. Harris, "Is Leptin the Parabiotic 'Satiety' Factor? Past and Present Interpretations," *Appetite* 61, no. 1 (February 2013): 111–18.

123 *accumulation of fat:* T. Rennie, "Obesity as a Manifestation of Personality Disturbance," *Diseases of the Central Nervous System* 1 (1940): 238.

123 *in the short term:* R. Neumann, "Experimentelle Beitrage zur Lehre von dem taglichen Nahrungsbedarf des Menschen unter besonderer Berucksichtigung der notwendigen Eiweissmenge," *Arch Für Hyg* 45 (1902): 1–87.

123 *World War II:* A. Keys, J. Brožek, A. Henschel, O. Mickelsen, and H. Longstreet Taylor, *The Biology of Human Starvation,* 2 vols. (Minneapolis, MN: University of Minnesota Press, 1950), 1385.

124 *four- to six-month period:* E. A. Sims, R. F. Goldman, C. M. Gluck, E. S. Horton, P. C. Kelleher, and D. W. Rowe, "Experimental Obesity in Man," *Transactions of the Association of American Physicians* 81 (1968): 153–70.

124   *most adult men require:* E. A. Sims and E. S. Horton, "Endocrine and Metabolic Adaptation to Obesity and Starvation," *American Journal of Clinical Nutrition* 21, no. 12 (December 1, 1968): 1455–70.

125   *role in the project:* Shell, *The Hungry Gene.*

125   *in the journal* Nature*:* Y. Zhang, R. Proenca, M. Maffei, M. Barone, L. Leopold, and J. M. Friedman, "Positional Cloning of the Mouse Obese Gene and Its Human Homologue," *Nature* 372, no. 6505 (December 1, 1994): 425–32.

125   *patent for leptin:* Shell, *The Hungry Gene.*

126   *exactly as predicted:* J. L. Halaas, K. S. Gajiwala, M. Maffei, S. L. Cohen, B. T. Chait, D. Rabinowitz, et al., "Weight-Reducing Effects of the Plasma Protein Encoded by the Obese Gene," *Science* 269, no. 5233 (July 28, 1995): 543–46; L. A. Campfield, F. J. Smith, Y. Guisez, R. Devos, and P. Burn, "Recombinant Mouse OB Protein: Evidence for a Peripheral Signal Linking Adiposity and Central Neural Networks," *Science* 269, no. 5223 (July 28, 1995): 546–49.

126   *Cambridge, in 1996:* C. T. Montague, I. S. Farooqi, J. P. Whitehead, M. A. Soos, H. Rau, N. J. Wareham, et al., "Congenital Leptin Deficiency Is Associated with Severe Early-Onset Obesity in Humans," *Nature* 387, no. 6636 (June 26, 1997): 903–908.

127   *seemingly insatiable appetites:* Shell, *The Hungry Gene.*

127   *inactivate the leptin gene:* Montague, Farooqi, Whitehead, Soos, Rau, Wareham, et al., "Congenital Leptin Deficiency."

127   *high intake of calories:* I. S. Farooqi and S. O'Rahilly, "Leptin: A Pivotal Regulator of Human Energy Homeostasis," *American Journal of Clinical Nutrition* 89, no. 3 (March 1, 2009): 980S–984S.

127   *high-reward foods:* Ibid.

128   *directly from the freezer:* Shell, *The Hungry Gene.*

128   *remarkable obsession with food:* Keys, Brožek, Henschel, Mickelsen, and Longstreet Taylor, *The Biology of Human Starvation.*

129   *slimmer body size:* R. L. Leibel and J. Hirsch, "Diminished Energy Requirements in Reduced-Obese Patients," *Metabolism* 33, no. 2 (February 1984): 164–70.

129   *experience after weight loss:* M. Rosenbaum, R. Goldsmith, D. Bloomfield, A. Magnano, L. Weimer, S. Heymsfield, et al., "Low-Dose Leptin Reverses Skeletal Muscle, Autonomic, and Neuroendocrine Adaptations to Maintenance of Reduced Weight," *Journal of Clinical Investigation* 115, no. 12 (December 2005): 3579–86.

129   *high-reward foods:* M. Rosenbaum, M. Sy, K. Pavlovich, R. L. Leibel, and J. Hirsch, "Leptin Reverses Weight Loss–Induced Changes in Regional Neural Activity Responses to Visual Food Stimuli," *Journal of Clinical Investigation* 118, no. 7 (July 1, 2008): 2583–91; H. R. Kissileff, J. C. Thornton, M. I. Torres, K. Pavlovich, L. S. Mayer, V. Kalari, et al., "Leptin Reverses Declines in Satiation in Weight-Reduced Obese Humans," *American Journal of Clinical Nutrition* 95, no. 2 (February 2012): 309–17.

129   *a helpful analogy:* J. M. Friedman, "A War on Obesity, Not the Obese," *Science* 299, no. 5608 (February 7, 2003): 856–58.

130   *before weight loss:* Rosenbaum, Goldsmith, Bloomfield, Magnano, Weimer, Heymsfield, et al., "Low-Dose Leptin"; Rosenbaum, Sy, Pavlovich, Leibel, and

Hirsch, "Leptin Reverses Weight Loss–Induced Changes"; Kissileff, Thornton, Torres, Pavlovich, Mayer, Kalari, et al., "Leptin Reverses Declines."

131 *tempting foods normalized:* I. S. Farooqi, E. Bullmore, J. Keogh, J. Gillard, S. O'Rahilly, and P. C. Fletcher, "Leptin Regulates Striatal Regions and Human Eating Behavior," *Science* 317, no. 5843 (September 7, 2007): 1355.

131 *similar to normal kids:* I. S. Farooqi, G. Matarese, G. M. Lord, J. M. Keogh, E. Lawrence, C. Agwu, et al., "Beneficial Effects of Leptin on Obesity, T Cell Hyporesponsiveness, and Neuroendocrine/Metabolic Dysfunction of Human Congenital Leptin Deficiency," *Journal of Clinical Investigation* 110, no. 8 (October 15, 2002): 1093–103.

131 *high levels of leptin:* R. V. Considine, M. K. Sinha, M. L. Heiman, A. Kriauciunas, T. W. Stephens, M. R. Nyce, et al., "Serum Immunoreactive Leptin Concentrations in Normal-weight and Obese Humans," *New England Journal of Medicine* 334, no. 5 (February 1996): 292–95.

131 *normal circulating amount:* S. B. Heymsfield, A. S. Greenberg, K. Fujioka, R. M. Dixon, R. Kushner, T. Hunt, et al., "Recombinant Leptin for Weight Loss in Obese and Lean Adults: A Randomized, Controlled, Dose-Escalation Trial," *Journal of the American Medical Association* 282, no. 16 (October 27, 1999): 1568–75.

132 *more than others:* Y. Ravussin, R. L. Leibel, and A. W. Ferrante Jr., "A Missing Link in Body Weight Homeostasis: The Catabolic Signal of the Overfed State," *Cell Metabolism* 20, no. 4 (October 7, 2014): 565–72.

134 *eating a smaller meal:* S. Carnell and J. Wardle, "Appetite and Adiposity in Children: Evidence for a Behavioral Susceptibility Theory of Obesity," *American Journal of Clinical Nutrition* 88, no. 1 (July 2008): 22–29.

135 *low-carbohydrate diets:* Group DPPR, "10-Year Follow-Up of Diabetes Incidence and Weight Loss in the Diabetes Prevention Program Outcomes Study," *Lancet* 374, no. 9702 (November 20, 2009): 1677–86; I. Shai, D. Schwarzfuchs, Y. Henkin, D. R. Shahar, S. Witkow, I. Greenberg, et al., "Weight Loss with a Low-Carbohydrate, Mediterranean, or Low-Fat Diet," *New England Journal of Medicine* 359, no. 3 (July 17, 2008): 229–41.

135 *against her hypothalamus:* "*Biggest Loser* Winner Admits She's Gained Back All the Weight," Mail Online, 2016, cited April 22, 2016, http://www.dailymail.co .uk/femail/article-3550257/I-feel-like-failure-Biggest-Loser-winner-dropped -112lbs-shares-shame-embarrassment-admitting-gained-nearly-weight-lost .html.

135 *we're all fat again:* "Former *Biggest Loser* Contestants Admit 'We Are All Fat Again!'" Mail Online, 2015, cited April 22, 2016, http://www.dailymail.co.uk /femail/article-2927207/We-fat-Former-Biggest-Loser-contestants-admit -controversial-regained-weight-endure-lasting-health-issues.html.

136 *the affluent world:* Stokes, "Using Maximum Weight"; Stokes and Preston, "Revealing the Burden of Obesity Using Weight Histories"; S. S. Du Plessis, S. Cabler, D. A. McAlister, E. Sabanegh, and A. Agarwal, "The Effect of Obesity on Sperm Disorders and Male Infertility," *Nature Reviews Urology* 7, no. 3

(March 2010): 153–61; C. J. Brewer and A. H. Balen, "The Adverse Effects of Obesity on Conception and Implantation," *Reproduction* (Cambridge, England) 140, no. 3 (September 2010): 347–64.

136 *effect in rats:* B. E. Levin and A. A. Dunn-Meynell, "Defense of Body Weight Against Chronic Caloric Restriction in Obesity-Prone and -Resistant Rats," *American Journal of Physiology—Regulatory, Integrative and Comparative Physiology* 278, no. 1 (January 2000): R231–R237.

137 *three different diets:* B. E. Levin and A. A. Dunn-Meynell, "Defense of Body Weight Depends on Dietary Composition and Palatability in Rats with Diet-Induced Obesity," *American Journal of Physiology—Regulatory, Integrative and Comparative Physiology* 282, no. 1 (January 2002): R46–R54.

138 *through a straw:* Hashim and Van Itallie, "Studies in Normal and Obese Subjects"; Campbell, Hashim and Van Itallie, "Studies of Food-Intake Regulation."

138 *expanded upon these findings:* Cabanac and Rabe, "Influence of a Monotonous Food."

140 *less weight over time:* K. H. Schmitz, D. R. Jacobs, A. S. Leon, P. J. Schreiner, and B. Sternfeld, "Physical Activity and Body Weight: Associations over Ten Years in the CARDIA Study. Coronary Artery Risk Development in Young Adults," *International Journal of Obesity and Related Metabolic Disorders* 24, no. 11 (November 2000): 1475–87; A. J. Littman, A. R. Kristal, and E. White, "Effects of Physical Activity Intensity, Frequency, and Activity Type on 10-Y Weight Change in Middle-Aged Men and Women," *International Journal of Obesity* 29, no. 5 (January 11, 2005): 524–33.

140 *rats on the same diet:* B. E. Levin and A. A. Dunn-Meynell, "Chronic Exercise Lowers the Defended Body Weight Gain and Adiposity in Diet-Induced Obese Rats," *American Journal of Physiology—Regulatory, Integrative and Comparative Physiology* 286, no. 4 (April 2004): R771–R778.

140 *face of overeating:* C. Bouchard, A. Tchernof, and A. Tremblay, "Predictors of Body Composition and Body Energy Changes in Response to Chronic Overfeeding," *International Journal of Obesity* 38, no. 2 (February 1, 2014): 236–42.

140 *hardly lose any weight:* W. C. Miller, D. M. Koceja, and E. J. Hamilton, "A Meta-Analysis of the Past 25 Years of Weight Loss Research Using Diet, Exercise or Diet Plus Exercise Intervention," *International Journal of Obesity and Related Metabolic Disorders* 21, no. 10 (October 1997): 941–47.

140 *the exercise regimen:* C. A. Slentz, B. D. Duscha, J. L. Johnson, K. Ketchum, L. B. Aiken, G. P. Samsa, et al., "Effects of the Amount of Exercise on Body Weight, Body Composition, and Measures of Central Obesity: STRRIDE—A Randomized Controlled Study," *Archives of Internal Medicine* 164, no. 1 (January 12, 2004): 31–39; N. A. King, M. Hopkins, P. Caudwell, R. J. Stubbs, and J. E. Blundell, "Individual Variability Following 12 Weeks of Supervised Exercise: Identification and Characterization of Compensation for Exercise-Induced Weight Loss," *International Journal of Obesity* 32, no. 1 (January 2008): 177–84; J. E. Donnelly, J. J. Honas, B. K. Smith, M. S. Mayo, C. A. Gibson, D. K. Sullivan, et al., "Aerobic Exercise Alone Results in Clinically Significant Weight Loss for Men and Women:

Midwest Exercise Trial 2," *Obesity* (Silver Spring, MD) 21, no. 3 (March 2013): E219–E228.

141 *for twelve weeks:* King, Hopkins, Caudwell, Stubbs, and Blundell, "Individual Variability Following 12 Weeks."

142 *periods of about a year:* M. Hession, C. Rolland, U. Kulkarni, A. Wise, and J. Broom, "Systematic Review of Randomized Controlled Trials of Low-Carbohydrate vs. Low-Fat/Low-Calorie Diets in the Management of Obesity and Its Comorbidities," *Obesity Reviews* 10, no. 1 (January 1, 2009): 36–50.

142 *research backs this up:* S. M. Nickols-Richardson, M. D. Coleman, J. J. Volpe, and K. W. Hosig, "Perceived Hunger Is Lower and Weight Loss Is Greater in Overweight Premenopausal Women Consuming a Low-Carbohydrate/High-Protein vs. High-Carbohydrate/Low-Fat Diet," *Journal of the American Dietetic Association* 105, no. 9 (September 2005): 1433–37.

142 *eat fewer calories:* B. J. Brehm, S. E. Spang, B. L. Lattin, R. J. Seeley, S. R. Daniels, and D. A. D'Alessio, "The Role of Energy Expenditure in the Differential Weight Loss in Obese Women on Low-Fat and Low-Carbohydrate Diets," *Journal of Clinical Endocrinology and Metabolism* 90, no. 3 (March 1, 2005): 1475–82.

142 *influencing the lipostat system:* E. R. Ropelle, J. R. Pauli, M. F. A. Fernandes, S. A. Rocco, R. M. Marin, J. Morari, et al., "A Central Role for Neuronal AMP-Activated Protein Kinase (AMPK) and Mammalian Target of Rapamycin (mTOR) in High-Protein Diet–Induced Weight Loss," *Diabetes* 57, no. 3 (March 1, 2008): 594–605.

143 *striking this effect can be:* D. S. Weigle, P. A. Breen, C. C. Matthys, H. S. Calla-han, K. E. Meeuws, V. R. Burden, et al., "A High-Protein Diet Induces Sustained Reductions in Appetite, Ad Libitum Caloric Intake, and Body Weight Despite Compensatory Changes in Diurnal Plasma Leptin and Ghrelin Concentrations," *American Journal of Clinical Nutrition* 82, no. 1 (July 2005): 41–48.

143 *often accompanies dieting:* M. S. Westerterp-Plantenga, S. G. Lemmens, and K. R. Westerterp, "Dietary Protein—Its Role in Satiety, Energetics, Weight Loss and Health," *British Journal of Nutrition* 108, supplement S2 (August 2012): S105–S112.

143 *low-carbohydrate diets:* S. Soenen, A. G. Bonomi, S. G. T. Lemmens, J. Scholte, M. A. Thijssen, F. van Berkum, et al., "Relatively High-Protein or 'Low-Carb' Energy-Restricted Diets for Body Weight Loss and Body Weight Maintenance?" *Physiology and Behavior* 107, no. 3 (October 10, 2012): 374–80.

7. The Hunger Neuron

146 *brain of a rat:* J. T. Clark, P. S. Kalra, W. R. Crowley, and S. P. Kalra, "Neuropeptide Y and Human Pancreatic Polypeptide Stimulate Feeding Behavior in Rats," *Endocrinology* 115, no. 1 (July 1984): 427–29.

146 *involved in hunger:* J. D. White and M. Kershaw, "Increased Hypothalamic Neuro-peptide Y Expression Following Food Deprivation," *Molecular and Cellular Neuroscience* 1, no. 1 (August 1990): 41–48.

146 *reduces food intake:* M. W. Schwartz, J. L. Marks, A. J. Sipolst, D. G. Baskin,

S. C. Woods, S. E. Kahn, et al., "Central Insulin Administration Reduces Neuropeptide Y mRNA Expression in the Arcuate Nucleus of Food-Deprived Lean (Fa/Fa) but Not Obese (fa/fa) *Zucker* Rats," *Endocrinology* 128, no. 5 (May 1, 1991): 2645–47.

146n *higher rate than lean people:* B. Mittendorfer, F. Magkos, E. Fabbrini, B. S. Mohammed, and S. Klein, "Relationship Between Body Fat Mass and Free Fatty Acid Kinetics in Men and Women," *Obesity* (Silver Spring, MD) 17, no. 10 (October 2009): 1872–77.

146n *isn't typically the case:* M.-F. Hivert, M.-F. Langlois, and A. C. Carpentier, "The Entero-Insular Axis and Adipose Tissue-Related Factors in the Prediction of Weight Gain in Humans," *International Journal of Obesity* 31, no. 5 (November 28, 2006): 731–42.

148 *Eli Lilly, had scooped him:* T. W. Stephens, M. Basinski, P. K. Bristow, J. M. Bue-Valleskey, S. G. Burgett, L. Craft, et al., "The Role of Neuropeptide Y in the Antiobesity Action of the Obese Gene Product," *Nature* 377, no. 6549 (October 12, 1995): 530–2.

148 *as visible as* Science: M. W. Schwartz, D. G. Baskin, T. R. Bukowski, J. L. Kuijper, D. Foster, G. Lasser, et al., "Specificity of Leptin Action on Elevated Blood Glucose Levels and Hypothalamic Neuropeptide Y Gene Expression in *Ob/Ob* Mice," *Diabetes* 45, no. 4 (April 1996): 531–35.

148 *receptor for leptin:* M. W. Schwartz, R. J. Seeley, L. A. Campfield, P. Burn, and D. G. Baskin, "Identification of Targets of Leptin Action in Rat Hypothalamus," *Journal of Clinical Investigation* 98, no. 5 (September 1, 1996): 1101–106.

148 *adiposity via the brain:* R. J. Seeley, K. A. Yagaloff, S. L. Fisher, P. Burn, T. E. Thiele, G. van Dijk, et al., "Melanocortin Receptors in Leptin Effects," *Nature* 390, no. 6658 (November 27, 1997): 349.

149 *normally behaving mice:* Y. Aponte, D. Atasoy, and S. M. Sternson, "AGRP Neurons Are Sufficient to Orchestrate Feeding Behavior Rapidly and Without Training," *Nature Neuroscience* 14, no. 3 (March 2011): 351–55.

149 *until it eats:* J. N. Betley, S. Xu, Z. F. H. Cao, R. Gong, C. J. Magnus, Y. Yu, et al., "Neurons for Hunger and Thirst Transmit a Negative-Valence Teaching Signal," *Nature* 521, no. 7551 (May 14, 2015): 180–85.

150 *distinguish from normal mice:* Q. Wu, B. B. Whiddon, and R. D. Palmiter, Ablation of Neurons Expressing Agouti-Related Protein, but Not Melanin Concentrating Hormone, in Leptin-Deficient Mice Restores Metabolic Functions and Fertility," *Proceedings of the National Academy of Sciences of the United States of America* 109, no. 8 (February 21, 2012): 3155–60.

150 *roles in the system:* S. J. Guyenet and M. W. Schwartz, "Regulation of Food Intake, Energy Balance, and Body Fat Mass: Implications for the Pathogenesis and Treatment of Obesity," *Journal of Clinical Endocrinology and Metabolism* 97, no. 3 (March 2012): 745–55.

152 *mouse were a marionette:* Aponte, Atasoy, and Sternson, "AGRP neurons Are Sufficient"; Betley, Xu, Cao, Gong, Magnus, Yu, et al., "Neurons for Hunger and Thirst"; Wu, Whiddon, and Palmiter, "Ablation of Neurons"; M. E. Carter, M. E. Soden, L. S. Zweifel, and R. D. Palmiter, "Genetic Identification of a Neural Circuit That Suppresses Appetite," *Nature* 503, no. 7474 (November 7, 2013):

111–14; B. P. Shah, L. Vong, D. P. Olson, S. Koda, M. J. Krashes, C. Ye, et al., "MC4R-Expressing Glutamatergic Neurons in the Paraventricular Hypothalamus Regulate Feeding and Are Synaptically Connected to the Parabrachial Nucleus," *Proceedings of the National Academy of Sciences of the United States of America* 111, no. 36 (September 9, 2014): 13193–98.

152 *countless unique tasks:* F. A. C. Azevedo, L. R. B. Carvalho, L. T. Grinberg, J. M. Farfel, R. E. L. Ferretti, and R. E. P. Leite, et al., "Equal Numbers of Neuronal and Nonneuronal Cells Make the Human Brain an Isometrically Scaled-Up Primate Brain," *Journal of Comparative Neurology* 513, no. 5 (April 10, 2009): 532–41.

153 *happening in the brain:* C. T. De Souza, E. P. Araujo, S. Bordin, R. Ashimine, R. L. Zollner, A. C. Boschero, et al., "Consumption of a Fat-Rich Diet Activates a Proinflammatory Response and Induces Insulin Resistance in the Hypothalamus," *Endocrinology* 146, no. 10 (October 1, 2005): 4192–99.

154 *what they observed:* Ibid.

154 *resistance and weight gain:* X. Zhang, G. Zhang, H. Zhang, M. Karin, H. Bai, and D. Cai, "Hypothalamic IKKbeta/NF-kappaB and ER Stress Link Overnutrition to Energy Imbalance and Obesity," *Cell* 135, no. 1 (October 3, 2008): 61–73.

154 *activity of the leptin receptor:* C. Bjørbæk, H. J. Lavery, S. H. Bates, R. K. Olson, S. M. Davis, J. S. Flier, et al., "SOCS3 Mediates Feedback Inhibition of the Leptin Receptor via Tyr985," *Journal of Biological Chemistry* 275, no. 51 (December 22, 2000): 40649–57.

154 *a fattening diet:* Zhang, Zhang, Zhang, Karin, Bai, and Cai, "Hypothalamic IKKbeta/NF-kappaB"; H. Mori, R. Hanada, T. Hanada, D. Aki, R. Mashima, H. Nishinakamura, et al., "Socs3 Deficiency in the Brain Elevates Leptin Sensitivity and Confers Resistance to Diet-Induced Obesity," *Nature Medicine* 10, no. 7 (July 2004): 739–43.

155 *development of obesity:* J. P. Thaler, C.-X. Yi, E. A. Schur, S. J. Guyenet, B. H. Hwang, M. O. Dietrich, et al., "Obesity Is Associated with Hypothalamic Injury in Rodents and Humans," *Journal of Clinical Investigation* 122, no. 1 (January 3, 2012): 153–62.

156 *he was to be obese:* Ibid.; E. A. Schur, S. J. Melhorn, S.-K. Oh, J. M. Lacy, K. E. Berkseth, S. J. Guyenet, et al., "Radiologic Evidence That Hypothalamic Gliosis Is Associated with Obesity and Insulin Resistance in Humans," *Obesity* (Silver Spring, MD) 23, no. 11 (November 2015): 2142–48.

157 *at least in mice:* K. E. Berkseth, S. J. Guyenet, S. J. Melhorn, D. Lee, J. P. Thaler, E. A. Schur, et al., "Hypothalamic Gliosis Associated with High-Fat Diet Feeding Is Reversible in Mice: A Combined Immunohistochemical and Magnetic Resonance Imaging Study," *Endocrinology* 155, no. 8 (August 2014): 2858–67.

157 *the gut microbiota:* P. D. Cani, J. Amar, M. A. Iglesias, M. Poggi, C. Knauf, D. Bastelica, et al., "Metabolic Endotoxemia Initiates Obesity and Insulin Resistance," *Diabetes* 56, no. 7 (July 2007): 1761–72.

157 *are less fattening:* S. C. Benoit, C. J. Kemp, C. F. Elias, W. Abplanalp, J. P. Herman, S. Migrenne, et al., "Palmitic Acid Mediates Hypothalamic Insulin

Resistance by Altering PKC-? Subcellular Localization in Rodents," *Journal of Clinical Investigation* 119, no. 9 (September 1, 2009): 2577–89.

157n *pathway that has been implicated:* L. Ozcan, A. S. Ergin, A. Lu, J. Chung, S. Sarkar, D. Nie, et al., "Endoplasmic Reticulum Stress Plays a Central Role in Development of Leptin Resistance," *Cell Metabolism* 9, no. 1 (January 7, 2009): 35–51.

157n *neurons in the hypothalamus:* J. Li, Y. Tang, and D. Cai, "IKKβ/NF-κB Disrupts Adult Hypothalamic Neural Stem Cells to Mediate Neurodegenerative Mechanism of Dietary Obesity and Pre-Diabetes," *Nature Cell Biology* 14, no. 10 (October 2012): 999–1012.

158 *holidays are over:* J. A. Yanovski, S. Z. Yanovski, K. N. Sovik, T. T. Nguyen, P. M. O'Neil, and N. G. Sebring, "A Prospective Study of Holiday Weight Gain," *New England Journal of Medicine* 342, no. 12 (March 23, 2000): 861–67.

158 *contribute to leptin resistance:* Z. A. Knight, K. S. Hannan, M. L. Greenberg, and J. M. Friedman, "Hyperleptinemia Is Required for the Development of Leptin Resistance," *PLOS ONE* 5, no. 6 (2010): e11376; K. M. Gamber, L. Huo, S. Ha, J. E. Hairston, S. Greeley, and C. Bjørbæk, "Over-Expression of Leptin Receptors in Hypothalamic POMC Neurons Increases Susceptibility to Diet-Induced Obesity," *PLOS ONE* 7, no. 1 (January 20, 2012): e30485; C. L. White, A. Whittington, M. J. Barnes, Z. Wang, G. A. Bray, and C. D. Morrison, "HF Diets Increase Hypothalamic PTP1B and Induce Leptin Resistance Through Both Leptin-Dependent and-Independent Mechanisms," *American Journal of Physiology—Endocrinology and Metabolism* 296, no. 2 (February 2009): E291–E299.

158 *changes in calorie intake:* C. Chin-Chance, K. S. Polonsky, and D. A. Schoeller, "Twenty-Four-Hour Leptin Levels Respond to Cumulative Short-Term Energy Imbalance and Predict Subsequent Intake," *Journal of Clinical Endocrinology and Metabolism* 85, no. 8 (August 2000): 2685–91.

159 *set point of the lipostat:* White, Whittington, Barnes, Wang, Bray, and Morrison, "HF Diets Increase Hypothalamic PTP1B"; Y. Ravussin, C. A. LeDuc, K. Watanabe, B. R. Mueller, A. Skowronski, M. Rosenbaum, et al., "Effects of Chronic Leptin Infusion on Subsequent Body Weight and Composition in Mice: Can Body Weight Set Point Be Reset?" *Molecular Metabolism* 3, no. 4 (March 5, 2014): 432–40.

159 *calorie-dense foods:* E. A. Schur, N. M. Kleinhans, J. Goldberg, D. Buchwald, M. W. Schwartz, and K. Maravilla, "Activation in Brain Energy Regulation and Reward Centers by Food Cues Varies with Choice of Visual Stimulus," *International Journal of Obesity* 33, no. 6 (April 14, 2009): 653–61.

162 *food cues subsides:* S. Mehta, S. J. Melhorn, A. Smeraglio, V. Tyagi, T. Grabowski, M. W. Schwartz, et al., "Regional Brain Response to Visual Food Cues Is a Marker of Satiety That Predicts Food Choice," *American Journal of Clinical Nutrition* 96, no. 5 (November 1, 2012): 989–99.

163 *refuse additional food:* H. J. Grill and R. Norgren, "Chronically Decerebrate Rats Demonstrate Satiation but Not Bait Shyness," *Science* 201, no. 4352 (July 21, 1978): 267–69.

164 *produces when we eat:* R. J. Seeley, H. J. Grill, and J. M. Kaplan, "Neurological

Dissociation of Gastrointestinal and Metabolic Contributions to Meal Size Control," *Behavioral Neuroscience* 108, no. 2 (1994): 347–52; H. J. Grill and G. P. Smith, "Cholecystokinin Decreases Sucrose Intake in Chronic Decerebrate Rats," *American Journal of Physiology—Regulatory, Integrative and Comparative Physiology* 254, no. 6 (June 1, 1988): R853–R856.

165 *act on the brain directly:* Guyenet and Schwartz, "Regulation of Food Intake."

165 *what you ate:* H. J. Grill and M. R. Hayes, "Hindbrain Neurons as an Essential Hub in the Neuroanatomically Distributed Control of Energy Balance," *Cell Metabolism* 16, no. 3 (September 5, 2012): 296–309.

166 *size of subsequent meals:* J. M. Kaplan, R. J. Seeley, and H. J. Grill, "Daily Caloric Intake in Intact and Chronic Decerebrate Rats," *Behavioral Neuroscience* 107, no. 5 (October 1993): 876–81.

166 *meal-to-meal food intake:* Grill and Hayes, "Hindbrain Neurons."

166 *NTS neurons:* Carter, Soden, Zweifel, and Palmiter, "Genetic Identification"; Shah, Vong, Olson, Koda, Krashes, Ye, et al., "MC4R-Expressing Glutamatergic Neurons."

166 *long-term changes in adiposity:* Guyenet and Schwartz, "Regulation of Food Intake."

167 *brains are leptin resistant:* A. S. Bruce, L. M. Holsen, R. J. Chambers, L. E. Martin, W. M. Brooks, J. R. Zarcone, et al., "Obese Children Show Hyperactivation to Food Pictures in Brain Networks Linked to Motivation, Reward and Cognitive Control," *International Journal of Obesity* 34, no. 10 (October 2010): 1494–500.

167 *full with fewer calories:* S. H. Holt, J. C. Miller, P. Petocz, and E. Farmakalidis, "A Satiety Index of Common Foods," *European Journal of Clinical Nutrition* 49, no. 9 (September 1995): 675–90.

168 *neurons in the LH:* T. H. Park and K. D. Carr, "Neuroanatomical Patterns of Fos-Like Immunoreactivity Induced by a Palatable Meal and Meal-Paired Environment in Saline- and Naltrexone-Treated Rats," *Brain Research* 805, nos. 1,2 (September 14, 1998): 169–80.

168 *make us feel full:* Ibid.; C. Jiang, R. Fogel, and X. Zhang, "Lateral Hypothalamus Modulates Gut-Sensitive Neurons in the Dorsal Vagal Complex," *Brain Research* 980, no. 1 (August 1, 2003): 31–47; E. M. Parise, N. Lilly, K. Kay, A. M. Dossat, R. Seth, J. M. Overton, et al., "Evidence for the Role of Hindbrain Orexin-1 Receptors in the Control of Meal Size," *American Journal of Physiology—Regulatory, Integrative and Comparative Physiology* 301, no. 6 (December 2011): R1692–R1699.

169 *high-carbohydrate foods:* B. J. Rolls, "The Role of Energy Density in the Overconsumption of Fat," *Journal of Nutrition* 130, 2S suppl. (February 2000): 268S–271S.

169 *per unit calorie:* S. D. Poppitt, D. McCormack, and R. Buffenstein, "Short-Term Effects of Macronutrient Preloads on Appetite and Energy Intake in Lean Women," *Physiology and Behavior* 64, no. 3 (June 1, 1998): 279–85.

169 *signal to the NTS:* R. Faipoux, D. Tomé, S. Gougis, N. Darcel, and G. Fromentin, "Proteins Activate Satiety-Related Neuronal Pathways in the Brainstem and Hypothalamus of Rats," *Journal of Nutrition* 138, no. 6 (June 1, 2008): 1172–78;

N. Geary, "Pancreatic Glucagon Signals Postprandial Satiety," *Neuroscience and Biobehaviorial Reviews* 14, no. 3 (1990): 323–38.

170   *lower calorie intake:* T. Jönsson, Y. Granfeldt, C. Erlanson-Albertsson, B. Ahrén, and S. Lindeberg, "A Paleolithic Diet Is More Satiating Per Calorie Than a Mediterranean-Like Diet in Individuals with Ischemic Heart Disease," *Nutrition and Metabolism* 7 (November 30, 2010): 85.

170   *explaining its popularity:* S. Lindeberg, T. Jönsson, Y. Granfeldt, E. Borgstrand, J. Soffman, K. Sjöström, et al., "A Palaeolithic Diet Improves Glucose Tolerance More Than a Mediterranean-Like Diet in Individuals with Ischaemic Heart Disease," *Diabetologia* 50, no. 9 (September 2007): 1795–807; T. Jönsson, Y. Granfeldt, B. Ahrén, U.-C. Branell, G. Pålsson, A. Hansson, et al., "Beneficial Effects of a Paleolithic Diet on Cardiovascular Risk Factors in Type 2 Diabetes: A Randomized Cross-Over Pilot Study," *Cardiovascular Diabetology* 8 (2009): 35.

171   *notions about obesity:* M. Börjeson, "The Aetiology of Obesity in Children. A Study of 101 Twin Pairs," *Acta Pædiatrica* 65, no. 3 (May 1, 1976): 279–87.

172   *body weight between individuals:* H. H. Maes, M. C. Neale, and L. J. Eaves, "Genetic and Environmental Factors in Relative Body Weight and Human Adiposity," *Behavior Genetics* 27, no. 4 (July 1997): 325–51.

172   *our food intake:* J. M. de Castro, "Genetic Influences on Daily Intake and Meal Patterns of Humans," *Physiology and Behavior* 53, no. 4 (April 1993): 777–82; J. M. de Castro, "Palatability and Intake Relationships in Free-Living Humans: The Influence of Heredity," *Nutrition Research* (New York) 21, no. 7 (July 2001): 935–45.

172   *for one hundred days:* C. Bouchard, A. Tremblay, J.-P. Després, A. Nadeau, P. J. Lupien, G. Thériault, et al., "The Response to Long-Term Overfeeding in Identical Twins," *New England Journal of Medicine* 322, no. 21 (May 24, 1990): 1477–82.

173   *"non-exercise activity thermogenesis" (NEAT):* J. A. Levine, N. L. Eberhardt, and M. D. Jensen, "Role of Nonexercise Activity Thermogenesis in Resistance to Fat Gain in Humans," *Science* 283, no. 5399 (January 8, 1999): 212–14.

173   *actions in the brain:* A. A. van der Klaauw and I. S. Farooqi, "The Hunger Genes: Pathways to Obesity," *Cell* 161, no. 1 (March 26, 2015): 119–32.

174   *melanocortins in the brain:* Ibid.

175   *adiposity between people:* A. E. Locke, B. Kahali, S. I. Berndt, A. E. Justice, T. H. Pers, F. R. Day, et al., "Genetic Studies of Body Mass Index Yield New Insights for Obesity Biology," *Nature* 518, no. 7538 (February 12, 2015): 197–206.

8. Rhythms

177   *one of two groups:* M.-P. St-Onge, A. L. Roberts, J. Chen, M. Kelleman, M. O'Keeffe, A. RoyChoudhury, et al., "Short Sleep Duration Increases Energy Intakes but Does Not Change Energy Expenditure in Normal-Weight Individuals," *American Journal of Clinical Nutrition* 94, no. 2 (August 2011): 410–16.

178   *unknown brain disorder:* L. C. Triarhou, "The Percipient Observations of Constantin von Economo on Encephalitis Lethargica and Sleep Disruption and Their Lasting

Impact on Contemporary Sleep Research," *Brain Research Bulletin* 69, no. 3 (April 14, 2006): 244–58.

179 *the area he identified:* C. B. Saper, T. E. Scammell, and J. Lu, "Hypothalamic Regulation of Sleep and Circadian Rhythms," *Nature* 437, no. 7063 (October 27, 2005): 1257–63.

179 *from the outside world:* Ibid.

180 *"flip-flop switch":* Ibid.

181 *adenosine is that signal:* Ibid.

181 *when we exert ourselves:* M. Dworak, P. Diel, S. Voss, W. Hollmann, and H. K. Strüder, "Intense Exercise Increases Adenosine Concentrations in Rat Brain: Implications for a Homeostatic Sleep Drive," *Neuroscience* 150, no. 4 (December 19, 2007): 789–95.

182 *scourges of aging:* L. Xie, H. Kang, Q. Xu, M. J. Chen, Y. Liao, M. Thiyagarajan, et al., "Sleep Drives Metabolite Clearance from the Adult Brain," *Science* 342, no. 6156 (October 18, 2013): 373–77.

182 *per night, for two weeks:* H. P. A. van Dongen, G. Maislin, J. M. Mullington, and D. F. Dinges, "The Cumulative Cost of Additional Wakefulness: Dose-Response Effects on Neurobehavioral Functions and Sleep Physiology from Chronic Sleep Restriction and Total Sleep Deprivation," *Sleep* 26, no. 2 (March 15, 2003): 117–26.

182 *familial fatal insomnia:* P. Montagna, P. Gambetti, P. Cortelli, E. Lugaresi, "Familial and Sporadic Fatal Insomnia," *Lancet Neurology* 2, no. 3 (March 2003): 167–76.

183 *sleep deprivation can kill:* A. Rechtschaffen, M. A. Gilliland, B. M. Bergmann, and J. B. Winter, "Physiological Correlates of Prolonged Sleep Deprivation in Rats," *Science* 221, no. 4606 (July 8, 1983): 182–84.

183 *pizza and doughnuts:* M. P. St-Onge, A. McReynolds, Z. B. Trivedi, A. L. Roberts, M. Sy, and J. Hirsch, "Sleep Restriction Leads to Increased Activation of Brain Regions Sensitive to Food Stimuli," *American Journal of Clinical Nutrition* 95, no. 4 (April 2012): 818–24; M. P. St-Onge, S. Wolfe, M. Sy, A. Shechter, and J. Hirsch, "Sleep Restriction Increases the Neuronal Response to Unhealthy Food in Normal-Weight Individuals," *International Journal of Obesity* 38, no. 3 (March 2014): 411–16.

183 *without even realizing it:* C. Benedict, S. J. Brooks, O. G. O'Daly, M. S. Almèn, A. Morell, K. Åberg, et al., "Acute Sleep Deprivation Enhances the Brain's Response to Hedonic Food Stimuli: An fMRI Study," *Journal of Clinical Endocrinology and Metabolism* 97, no. 3 (March 2012): E443–E447.

184 *100 Calories per day:* A. Shechter, R. Rising, J. B. Albu, and M. P. St-Onge, "Experimental Sleep Curtailment Causes Wake-Dependent Increases in 24-H Energy Expenditure as Measured by Whole-Room Indirect Calorimetry," *American Journal of Clinical Nutrition* 98, no. 6 (December 1, 2013): 1433–39.

184 *nine hours per night:* S. R. Patel and F. B. Hu, "Short Sleep Duration and Weight Gain: A Systematic Review," *Obesity* 16, no. 3 (2008): 643–53.

184 *22 percent in 1985:* E. S. Ford, T. J. Cunningham, and J. B. Croft, "Trends in

Self-Reported Sleep Duration among US Adults from 1985 to 2012," *Sleep* 38, no. 5 (May 2015): 829–32.

184   *trends among adolescents:* K. M. Keyes, J. Maslowsky, A. Hamilton, and J. Schulenberg, "The Great Sleep Recession: Changes in Sleep Duration among US Adolescents, 1991–2012," *Pediatrics* 135, no. 3 (March 2015): 460–68.

184   *likely to die overall:* F. P. Cappuccio, D. Cooper, L. D'Elia, P. Strazzullo, and M. A. Miller, "Sleep Duration Predicts Cardiovascular Outcomes: A Systematic Review and Meta-Analysis of Prospective Studies," *European Heart Journal* 32, no. 12 (June 2011): 1484–92; F. P. Cappuccio, L. D'Elia, P. Strazzullo, and M. A. Miller, "Sleep Duration and All-Cause Mortality: A Systematic Review and Meta-Analysis of Prospective Studies," *Sleep* 33, no. 5 (May 2010): 585–92; N. T. Ayas, D. P. White, W. K. Al-Delaimy, J. E. Manson, M. J. Stampfer, F. E. Speizer, et al., "A Prospective Study of Self-Reported Sleep Duration and Incident Diabetes in Women," *Diabetes Care* 26, no. 2 (February 1, 2003): 380–84.

185   *it has declined slightly:* Ford, Cunningham, and Croft, "Trends in Self-Reported Sleep Duration."

185   *such as sleep apnea:* P. E. Peppard, T. Young, J. H. Barnet, M. Palta, E. W. Hagen, and K. M. Hla, "Increased Prevalence of Sleep-Disordered Breathing in Adults," *American Journal of Epidemiology* 177, no. 9 (May 1, 2013): 1006–14.

186   *decision-making behavior:* V. Venkatraman, S. A. Huettel, L. Y. M. Chuah, J. W. Payne, M. W. L. Chee, "Sleep Deprivation Biases the Neural Mechanisms Underlying Economic Preferences," *Journal of Neuroscience* 31, no. 10 (March 9, 2011): 3712–18.

187   *night preceding the experiment:* D. Pardi, M. Buman, J. Black, G. Lammers, and J. Zeitzer, "Eating Decisions Based on Alertness Levels After a Single Night of Sleep Manipulation," In review.

187   *drive food reward:* S. M. Greer, A. N. Goldstein, and M. P. Walker, "The Impact of Sleep Deprivation on Food Desire in the Human Brain," *Nature Communications* 4 (August 6, 2013): 2259.

188   *French-Italian Maritime Alps:* M. Siffre, *Expériences Hors du Temps* (Paris, France: Fayard, 1972).

189   *every one of your cells:* Saper, Scammell, and Lu, "Hypothalamic Regulation of Sleep"; E. Bianconi, A. Piovesan, F. Facchin, A. Beraudi, R. Casadei, F. Frabetti, et al., "An Estimation of the Number of Cells in the Human Body," *Annals of Human Biology* 40, no. 6 (December 2013): 463–71.

189   *suprachiasmatic nucleus (SCN):* Saper, Scammell, and Lu, "Hypothalamic Regulation of Sleep."

189   *abundant at midday:* G. C. Brainard, J. P. Hanifin, J. M. Greeson, B. Byrne, G. Glickman, and E. Gerner, et al., "Action Spectrum for Melatonin Regulation in Humans: Evidence for a Novel Circadian Photoreceptor," *Journal of Neuroscience* 21, no. 16 (August 15, 2001): 6405–12.

189   *metabolism, and eating:* Saper, Scammell, and Lu, "Hypothalamic Regulation of Sleep."

190   *should be increasing:* I. M. McIntyre, T. R. Norman, G. D. Burrows, and S. M. Armstrong, "Human Melatonin Suppression by Light Is Intensity Dependent,"

Journal of Pineal Research 6, no. 2 (April 1, 1989): 149–56; A.-M. Chang, D. Aeschbach, J. F. Duffy, and C. A. Czeisler, "Evening Use of Light-Emitting eReaders Negatively Affects Sleep, Circadian Timing, and Next-Morning Alertness," *Proceedings of the National Academy of Sciences of the United States of America* 112, no. 4 (January 27, 2015): 1232–37.

191   *understand that it's nighttime:* L. Kayumov, R. F. Casper, R. J. Hawa, B. Perelman, S. A. Chung, S. Sokalsky, et al., "Blocking Low-wavelength Light Presents Nocturnal Melatonin Suppression With No Adverse Effects on Performance During Simulated Shift Work," *Journal of Clinical Endocrinology and Metabolism* 90, no. 5 (May 2005): 2755–61.

193   *cancer, and cardiovascular disease:* A. Pan, E. S. Schernhammer, Q. Sun, and F. B. Hu, "Rotating Night Shift Work and Risk of Type 2 Diabetes: Two Prospective Cohort Studies in Women," *PLOS Medicine* 8, no. 12 (December 6, 2011): e1001141; L. G. van Amelsvoort, E. G. Schouten, and F. J. Kok, "Duration of Shiftwork Related to Body Mass Index and Waist to Hip Ratio," *International Journal of Obesity and Related Metabolic Disorders* 23, no. 9 (September 1999): 973–78; E. S. Schernhammer, F. Laden, F. E. Speizer, W. C. Willett, D. J. Hunter, I. Kawachi, et al., "Rotating Night Shifts and Risk of Breast Cancer in Women Participating in the Nurses' Health Study," *Journal of the National Cancer Institute* 93, no. 20 (October 17, 2001): 1563–68; M. V. Vyas, A. X. Garg, A. V. Iansavichus, J. Costella, A. Donner, L. E. Laugsand, et al., "Shift Work and Vascular Events: Systematic Review and Meta-Analysis," *BMJ* 345 (July 26, 2012): e4800.

193   *same fattening diet:* D. M. Arble, J. Bass, A. D. Laposky, M. H. Vitaterna, and F. W. Turek, "Circadian Timing of Food Intake Contributes to Weight Gain," *Obesity* (Silver Spring, MD) 17, no. 11 (November 2009): 2100–102.

194   *Arble and Turek's findings:* M. Hatori, C. Vollmers, A. Zarrinpar, L. DiTacchio, E. A. Bushong, S. Gill, et al., "Time-Restricted Feeding Without Reducing Caloric Intake Prevents Metabolic Diseases in Mice Fed a High-Fat Diet," *Cell Metabolism* 15, no. 6 (June 6, 2012): 848–60; H. Sherman, Y. Genzer, R. Cohen, N. Chapnik, Z. Madar, and O. Froy, "Timed High-Fat Diet Resets Circadian Metabolism and Prevents Obesity," *FASEB Journal* 26, no. 8 (August 1, 2012): 3493–502; L. K. Fonken, J. L. Workman, J. C. Walton, Z. M. Weil, J. S. Morris, A. Haim, et al., "Light at Night Increases Body Mass by Shifting the Time of Food Intake," *Proceedings of the National Academy of Sciences of the United States of America* 107, no. 43 (October 26, 2010): 18664–69; I. N. Karatsoreos, S. Bhagat, E. B. Bloss, J. H. Morrison, and B. S. McEwen, "Disruption of Circadian Clocks Has Ramifications for Metabolism, Brain, and Behavior," *Proceedings of the National Academy of Sciences of the United States of America* 108, no. 4 (January 25, 2011): 1657–62.

194   *heavier than average:* S. L. Colles, J. B. Dixon, and P. E. O'Brien, "Night Eating Syndrome and Nocturnal Snacking: Association with Obesity, Binge Eating and Psychological Distress," *International Journal of Obesity* 31, no. 11 (June 19, 2007): 1722–30.

9. Life in the Fast Lane

198 *insomnia, and irritability:* American Psychological Association, *Stress in America* (2007).

198 *sometimes even predation:* Marlowe, *The Hadza*; N. A. Chagnon and E. O. Wilson, *Yanomamö: The Last Days of Eden* (San Diego: Jovanovich, 1992).

198 *stress in their lives:* APA, *Stress in America.*

198 *food intake, and weight loss:* D. D. Krahn, B. A. Gosnell, M. Grace, and A. S. Levine, "CRF Antagonist Partially Reverses CRF- and Stress-Induced Effects on Feeding," *Brain Research Bulletin* 17, no. 3 (September 1986): 285–89.

198 *confirmed by other studies:* T. C. Adam and E. S. Epel, "Stress, Eating and the Reward System," *Physiology and Behavior* 91, no. 4 (July 24, 2007): 449–58.

199 *threat response system:* H. Klüver and P. C. Bucy, "Preliminary Analysis of Functions of the Temporal Lobes in Monkeys," *Archives of Neurology and Psychiatry* 42, no. 6 (1939): 979–1000.

200 *sustain severe injuries:* M. Davis and P. J. Whalen, "The Amygdala: Vigilance and Emotion," *Molecular Psychiatry* 6, no. 1 (January 2001): 13–34.

200 *adjacent brain tissue:* Ibid.

201 *threat response system:* Ibid.; Y. M. Ulrich-Lai and J. P. Herman, "Neural Regulation of Endocrine and Autonomic Stress Responses," *Nature Reviews Neuroscience* 10, no. 6 (June 2009): 397–409; J. LeDoux, *Anxious: Using the Brain to Understand and Treat Fear and Anxiety,* 1st ed. (New York: Viking, 2015), 480.

201 *we commonly call "stress":* LeDoux, *Anxious.*

201 *signs of a threat:* Ibid.

202 *consequences of the situation:* Davis and Whalen, "The Amygdala"; LeDoux, *Anxious.*

202 *takes to resolve it:* Davis and Whalen, "The Amygdala."

202 *closing your eyes:* Ibid.; LeDoux, *Anxious.*

202 *your sympathetic nervous system:* Davis and Whalen, "The Amygdala"; Ulrich-Lai and Herman, "Neural Regulation"; LeDoux, *Anxious.*

203 *hypothalamic-pituitary-adrenal axis (HPA axis):* Ibid.

204 *stress hormone cortisol:* Ulrich-Lai and Herman, "Neural Regulation"; LeDoux, *Anxious.*

205 *pay the bills:* K. E. Habib, K. P. Weld, K. C. Rice, J. Pushkas, M. Champoux, S. Listwak, et al., "Oral Administration of a Corticotropin-Releasing Hormone Receptor Antagonist Significantly Attenuates Behavioral, Neuroendocrine, and Autonomic Responses to Stress in Primates," *Proceedings of the National Academy of Sciences of the United States of America* 97, no. 11 (May 23, 2000): 6079–84.

205 *effectively in the future:* E. A. Phelps and J. E. LeDoux, "Contributions of the Amygdala to Emotion Processing: From Animal Models to Human Behavior," *Neuron* 48, no. 2 (October 20, 2005): 175–87.

206 *believe we can control:* A. Breier, M. Albus, D. Pickar, T. P. Zahn, O. M. Wolkowitz, and S. M. Paul, "Controllable and Uncontrollable Stress in Humans: Alterations in Mood and Neuroendocrine and Psychophysiological Function," *American Journal of Psychiatry* 144, no. 11 (November 1987): 1419–25.

206 *monkeys maintain weight:* V. Michopoulos, M. Higgins, D. Toufexis, and M. E. Wilson, "Social Subordination Produces Distinct Stress-Related Phenotypes in Female Rhesus Monkeys," *Psychoneuroendocrinology* 37, no. 7 (July 2012): 1071–85.

207 *behavior changes dramatically:* V. Michopoulos, D. Toufexis, and M. E. Wilson, "Social Stress Interacts with Diet History to Promote Emotional Feeding in Females," *Psychoneuroendocrinology* 37, no. 9 (September 2012): 1479–90.

207 *monkeys stop overeating:* C. J. Moore, Z. P. Johnson, M. Higgins, D. Toufexis, and M. E. Wilson, "Antagonism of Corticotrophin-Releasing Factor Type 1 Receptors Attenuates Caloric Intake of Free Feeding Subordinate Female Rhesus Monkeys in a Rich Dietary Environment," *Journal of Neuroendocrinology* 27, no. 1 (January 2015): 33–43.

208 *had abdominal obesity:* H. Cushing, "The Basophil Adenomas of the Pituitary Body and Their Clinical Manifestations. Pituitary Basophilism," *Bulletin of the Johns Hopkins Hospital* L (1932): 137–95.

208 *because of such tumors:* L. K. Nieman and I. Ilias, "Evaluation and Treatment of Cushing's Syndrome," *American Journal of Medicine* 118, no. 12 (December 2005): 1340–46.

208 *each group's food intake:* P. A. Tataranni, D. E. Larson, S. Snitker, J. B. Young, J. P. Flatt, and E. Ravussin, "Effects of Glucocorticoids on Energy Metabolism and Food Intake in Humans," *American Journal of Physiology* 271, no. 2 pt. 1 (August 1996): E317–E325.

209 *rodent equivalent of cortisol:* K. E. Zakrzewska, I. Cusin, A. Sainsbury, F. Rohner-Jeanrenaud, and B. Jeanrenaud, "Glucocorticoids as Counterregulatory Hormones of Leptin: Toward an Understanding of Leptin Resistance," *Diabetes* 46, no. 4 (April 1, 1997): 717–19.

209 *hunger-promoting substance NPY:* R. Ishida-Takahashi, S. Uotani, T. Abe, M. Degawa-Yamauchi, T. Fukushima, N. Fujita, et al., "Rapid Inhibition of Leptin Signaling by Glucocorticoids In Vitro and In Vivo," *Journal of Biological Chemistry* 279, no. 19 (May 7, 2004): 19658–64; A. M. Strack, R. J. Sebastian, M. W. Schwartz, and M. F. Dallman, "Glucocorticoids and Insulin: Reciprocal Signals for Energy Balance," *American Journal of Physiology* 268, no. 1 pt. 2 (January 1995): R142–R149.

210 *reminiscent of Cushing's:* E. J. Brunner, T. Chandola, and M. G. Marmot, "Prospective Effect of Job Strain on General and Central Obesity in the Whitehall II Study," *American Journal of Epidemiology* 165, no. 7 (April 1, 2007): 828–37; J. P. Block, Y. He, A. M. Zaslavsky, L. Ding, and J. Z. Ayanian, "Psychosocial Stress and Change in Weight Among US Adults," *American Journal of Epidemiology* (January 1, 2009), doi: 10.1093/aje/kwp104; A. Kouvonen, M. Kivimäki, S. J. Cox, T. Cox, and J. Vahtera, "Relationship Between Work Stress and Body Mass Index Among 45,810 Female and Male Employees," *Psychosomatic Medicine* 67, no. 4 (August 2005): 577–83; T. Chandola, E. Brunner, and M. Marmot, "Chronic Stress at Work and the Metabolic Syndrome: Prospective Study," *BMJ* 332, no. 7540 (March 2, 2006): 521–25.

210 *much cortisol at all:* E. Epel, R. Lapidus, B. McEwen, and K. Brownell, "Stress May Add Bite to Appetite in Women: A Laboratory Study of Stress-Induced Cortisol and Eating Behavior," *Psychoneuroendocrinology* 26, no. 1 (January 2001): 37–49; E. Newman, D. B. O'Connor, and M. Conner, "Daily Hassles and Eating Behaviour: The Role of Cortisol Reactivity Status," *Psychoneuroendocrinology* 32, no. 2 (February 2007): 125–32.

210 *cortisol-raising effect:* Breier, Albus, Pickar, Zahn, Wolkowitz, and Paul, "Controllable and Uncontrollable Stress."

212 *types of food we eat:* APA, *Stress in America*; D. A. Zellner, S. Loaiza, Z. Gonzalez, J. Pita, J. Morales, D. Pecora, et al., "Food Selection Changes Under Stress," *Physiology and Behavior* 87, no. 4 (April 15, 2006): 789–93; G. Oliver, J. Wardle, and E. L. Gibson, "Stress and Food Choice: A Laboratory Study," *Psychosomatic Medicine* 62, no. 6 (December 2000): 853–65.

212 *got only plain water:* A. M. Strack, S. F. Akana, C. J. Horsley, and M. F. Dallman, "A Hypercaloric Load Induces Thermogenesis but Inhibits Stress Responses in the SNS and HPA System," *American Journal of Physiology* 272, no. 3 pt. 2 (March 1997): R840–R848.

212 *does the same thing:* S. E. la Fleur, H. Houshyar, M. Roy, and M. F. Dallman, "Choice of Lard, but Not Total Lard Calories, Damps Adrenocorticotropin Responses to Restraint," *Endocrinology* 146, no. 5 (May 2005): 2193–99.

213 *following restraint stress:* Y. M. Ulrich-Lai, M. M. Ostrander, I. M. Thomas, B. A. Packard, A. R. Furay, C. M. Dolgas, et al., "Daily Limited Access to Sweetened Drink Attenuates Hypothalamic-Pituitary-Adrenocortical Axis Stress Responses," *Endocrinology* 148, no. 4 (April 1, 2007): 1823–34.

213 sweet taste itself *was responsible:* Y. M. Ulrich-Lai, A. M. Christiansen, M. M. Ostrander, A. A. Jones, K. R. Jones, D. C. Choi, et al., "Pleasurable Behaviors Reduce Stress via Brain Reward Pathways," *Proceedings of the National Academy of Sciences of the United States of America* 107, no. 47 (November 23, 2010): 20529–34.

213 *tasty food: sex:* Ibid.

213 *stress-related information:* Ibid.

10. The Human Computer

221 *much less effort:* V. L. Gloy, M. Briel, D. L. Bhatt, S. R. Kashyap, P. R. Schauer, G. Mingrone, et al., "Bariatric Surgery versus Non-Surgical Treatment for Obesity: A Systematic Review and Meta-Analysis of Randomised Controlled Trials," *BMJ* 347 (October 22, 2013): f5934.

221 *they did before surgery:* K. A. Carswell, R. P. Vincent, A. P. Belgaumkar, R. A. Sherwood, S. A. Amiel, A. G. Patel, et al., "The Effect of Bariatric Surgery on Intestinal Absorption and Transit Time," *Obesity Surgery* 24, no. 5 (December 30, 2013): 796–805; E. A. Odstrcil, J. G. Martinez, C. A. S. Ana, B. Xue, R. E. Schneider, K. J. Steffer, et al., "The Contribution of Malabsorption to the Reduction in Net Energy Absorption after Long-Limb Roux-en-Y Gastric Bypass," *American Journal of Clinical Nutrition* 92, no. 4 (October 1, 2010): 704–13.

221 *vegetables and fruit:* C. N. Ochner, Y. Kwok, E. Conceição, S. P. Pantazatos, L. M.
     Puma, S. Carnell, et al., "Selective Reduction in Neural Responses to High Calorie
     Foods Following Gastric Bypass Surgery," *Annals of Surgery* 253, no. 3
     (March 2011): 502–507; A. D. Miras, R. N. Jackson, S. N. Jackson, A. P. Goldstone,
     T. Olbers, T. Hackenberg, et al., "Gastric Bypass Surgery for Obesity Decreases the
     Reward Value of a Sweet-Fat Stimulus as Assessed in a Progressive Ratio Task,"
     American Journal of Clinical Nutrition 96, no. 3 (September 1, 2012): 467–73;
     C. W. le Roux, M. Bueter, N. Theis, M. Werling, H. Ashrafian, C. Löwenstein,
     et al., "Gastric Bypass Reduces Fat Intake and Preference," *American Journal of
     Physiology—Regulatory, Integrative and Comparative Physiology* 301, no. 4 (October
     1, 2011): R1057–R1066; H. A. Kenler, R. E. Brolin, and R. P. Cody, "Changes in
     Eating Behavior After Horizontal Gastroplasty and Roux-en-Y Gastric Bypass,"
     American Journal of Clinical Nutrition 52, no. 1 (July 1, 1990): 87–92.
221 *undergone the same procedures:* le Roux, Bueter, Theis, Werling, Ashrafian,
     Löwenstein, et al., "Gastric Bypass"; H. Zheng, A. C. Shin, N. R. Lenard, R. L.
     Townsend, L. M. Patterson, D. L. Sigalet, et al., "Meal Patterns, Satiety, and Food
     Choice in a Rat Model of Roux-en-Y Gastric Bypass Surgery," *American Journal of
     Physiology—Regulatory, Integrative and Comparative Physiology* 297, no. 5 (Novem-
     ber 1, 2009): R1273–R1282; H. E. Wilson-Pérez, A. P. Chambers, D. A. Sandoval,
     M. A. Stefater, S. C. Woods, S. C. Benoit, et al., "The Effect of Vertical Sleeve
     Gastrectomy on Food Choice in Rats," *International Journal of Obesity* 37, no. 2
     (February 2013): 288–95.

11. Outsmarting the Hungry Brain

224 *nearly 10 percent:* Hall, Sacks, Chandramohan, Chow, Wang, Gortmaker, et al.,
     "Quantification of the Effect of Energy Imbalance."
224 *intake in real life:* B. Elbel, R. Kersh, V. L. Brescoll, and L. B. Dixon, "Calorie
     Labeling and Food Choices: A First Look at the Effects on Low-Income People in
     New York City," *Health Affairs* (Millwood) 28, no. 6 (November 1, 2009): w1110–
     w1121; B. Elbel, J. Gyamfi, and R. Kersh, "Child and Adolescent Fast-Food Choice
     and the Influence of Calorie Labeling: A Natural Experiment," *International
     Journal of Obesity* 35, no. 4 (April 2011): 493–500; J. Cantor, A. Torres, C. Abrams,
     and B. Elbel, "Five Years Later: Awareness of New York City's Calorie Labels
     Declined, with No Changes in Calories Purchased," *Health Affairs* (Millwood) 34,
     no. 11 (November 1, 2015): 1893–900.
225 *we did in 1963:* B. Forey, J. Hamling, J. Hamling, A. Thornton, and P. Lee,
     "Chapter 28: USA" in *International Smoking Statistics: A Collection of Historical
     Data from 30 Economically Developed Countries*, 2nd ed. (Oxford, UK: Oxford
     University Press, 2012).
225 *fattest countries in the world:* T. Rosenberg, "How One of the Most Obese Countries
     on Earth Took on the Soda Giants," *Guardian,* November 3, 2015, cited November
     9, 2015, http://www.theguardian.com/news/2015/nov/03/obese-soda-sugar-tax
     -mexico.

225 *years after being implemented:* D. Agren, "Mexico's Congress Accused of Caving to Soda Pop Industry in Tax Cut Plan," *Guardian,* October 19, 2015, cited November 4, 2015, http://www.theguardian.com/global-development/2015/oct/19/mexico-soda -tax-cut-pop-fizzy-drinks.

226 *Coca-Cola Mexico:* Rosenberg, "How One of the Most Obese Countries."

226 *$10 billion per year:* Associated Press, "Farm Subsidies Not in Sync with Food Pyramid," msnbc.com, cited November 9, 2015, http://www.nbcnews.com/id /8904252/ns/health-fitness/t/farm-subsidies-not-sync-food-pyramid/.

226 *two football fields:* Sections 92.0379A(j), B(j) & C(j) and 94.0379D(i) in City of Detroit, Official Zoning Ordinance.

227 *calorie-dense items:* Dembek, Harris, and Schwartz, "Where Children and Adolescents."

227 *government step in:* S. Speers, J. Harris, A. Goren, M. B. Schwartz, and K. D. Brownell, *Public Perceptions of Food Marketing to Youth: Results of the Rudd Center Public Opinion Poll, May 2008* (Rudd Center for Food Policy and Obesity, 2009).

227 *Coca-Cola, and General Mills:* E. D. Kolish, M. Enright, and B. Oberdorff, *The Children's Food & Beverage Advertising Initiative in Action: A Report on Compliance and Progress During 2013* (Children's Food & Beverage Advertising Initiative, 2014).

227 *comply with them reasonably well:* Ibid.

228 *arenas are unrestricted:* J. L. Harris, M. B. Schwartz, C. Shehan, M. Hyary, J. Appel, K. Haraghey, et al., *Snack F.A.C.T.S 2015: Evaluating Snack Food Nutrition and Marketing to Youth* (Rudd Center for Food Policy and Obesity, 2015).

228 *interest of public health:* Wansink, *Mindless Eating.*

228 *boycotts and international regulation:* M. Nestle, *Food Politics: How the Food Industry Influences Nutrition and Health,* revised and expanded edition (Berkeley: University of California Press, 2007), 510.

228 *"Monster Thickburger":* "Hardee's Serves Up 1,420-Calorie Burger," MSNBC, November 17, 2004, http://www.nbcnews.com/id/6498304/ns/business-us_business /t/hardees-serves—calorie-burger/.

229 *colossal hamburger:* "Hardee's Hails Burger as 'Monument to Decadence,'" *USA Today,* November 15, 2004, http://usatoday30.usatoday.com/money/industries/food /2004-11-15-hardees_x.htm.

230 *result of improper nutrition:* Stokes, "Using Maximum Weight"; A. Must and R. S. Strauss, "Risks and Consequences of Childhood and Adolescent Obesity," *International Journal of Obesity and Related Metabolic Disorders* 23, suppl. 2 (March 1999): S2–S11; A. Berrington de Gonzalez, P. Hartge, J. R. Cerhan, A. J. Flint, L. Hannan, R. J. MacInnis, et al., "Body-Mass Index and Mortality Among 1.46 Million White Adults," *New England Journal of Medicine* 363, no. 23 (2010): 2211–19.

232 *digestive system unabsorbed:* R. D. Mattes, P. M. Kris-Etherton, and G. D. Foster, "Impact of Peanuts and Tree Nuts on Body Weight and Healthy Weight Loss in Adults," *Journal of Nutrition* 138, no. 9 (September 2008): 1741S–1745S.

234 *does vary by individual:* King, Hopkins, Caudwell, Stubbs, and Blundell, "Individual Variability Following 12 Weeks"; D. M. Thomas, C. Bouchard, T.

Church, C. Slentz, W. E. Kraus, L. M. Redman, et al., "Why Do Individuals Not Lose More Weight from an Exercise Intervention at a Defined Dose? An Energy Balance Analysis," *Obesity Reviews* 13, no. 10 (October 2012): 835–47.

234 *lifting weights:* "2008 Physical Activity Guidelines for Americans," US Department of Health and Human Services, 2008, http://health.gov/paguidelines/pdf /paguide.pdf.

236 *can also improve health:* P. Grossman, L. Niemann, S. Schmidt, and H. Walach, "Mindfulness-Based Stress Reduction and Health Benefits: A Meta-Analysis," *Journal of Psychosomatic Research* 57, no. 1 (July 2004): 35–43; J. Daubenmier, J. Kristeller, F. M. Hecht, N. Maninger, M. Kuwata, K. Jhaveri, et al., "Mindfulness Intervention for Stress Eating to Reduce Cortisol and Abdominal Fat Among Overweight and Obese Women: An Exploratory Randomized Controlled Study," *Journal of Obesity* 2011 (2011): 651936.

Index

Page numbers followed by "f" indicate information followed by figures and page numbers followed by "n" indicate information found in notes.

A
abulia, 39–40, 39n
Abyss of Scarasson, 188
Aché hunter-gatherers, 94, 95
ACTH. *See* Adrenocorticotropic hormone
adaptation, 31
addiction
    avoidance and, 57
    delay discounting and, 109–110n
    impulsivity and, 67–68
    reinforcement and, 53–56
    withdrawal and, 53n
adenosine, sleep and, 181–182
adipose tissue, 12. *See also* Fat
adiposity, 12, 14–15
adiposity center, 166
adiposity factor, 166
adrenocorticotropic hormone (ACTH), 204f, 208
advertising
    countermarketing and, 224–225
    cravings and, 85–86
    principles of, 85–86n
    regulation of, 227–228
aggression, noncontact, 206
agouti mice, 118
agouti-related peptide (AgRP), 149n

AgRP neurons, 149n, 150n
akinesia, 38
alcohol
    caloric content of, 80–81, 81n
    reinforcement and, 49n, 80–81
    reward system and, 232
allelic effects, 174–175
Alzheimer's disease, 182
Amgen, 126
amygdala, threat response system and, 200–205, 201f
amyloid-β protein, 182
antalarmin, 205n
apnea, 185, 233
appetite management, 231–232
Arble, Deanna, 192–194
archery, 88n
arcuate nucleus, 146–148, 147f
arousal system
    brain regions of, 179, 180f
    encephalitis lethargica and, 178–179
artificial flavors, 80
astrocytes, 154–155, 154n, 155f
Atwater, Wilbur, 11–13
Audrain-McGovern, Janet, 109–110n
automobile ownership, 17, 18f
avoidance, role of, 57
awareness, 36n, 153

B
Baars, Ton, 2n
bacteria, chemotaxis and, 25n
bakery products, 167–168
baobab fruit, 91f
basal forebrain, 180f
basal ganglia
    human, 30
    lamprey, 26–28, 28f
    nonfunctional, 36–40
    reinforcement and, 44
    reward system and, 217
    roots of, 24n
    stress and, 205
    subjective value and, 103–105
beans, 167–168
behavior, brain as computer and, 216
Berridge, Kent, 44n
Berthoud, Hans-Rudi, 221
The Biggest Loser (TV show), 135
bilateral temporal monkey experiments, 200
Birch, Leann, 51
bland diet, 58–60, 138
bliss point, 77
blood oxygen, functional magnetic
    resonance imaging and, 160n
blue light, artificial, 190, 191–192, 233
Blundell, John, 141
body fat. See Fat (body)
body mass index (BMI), 9, 72n
body size, impacts of changes in on
    metabolic rate, 15, 15n
Börjeson, Mats, 171
Botswana, 70–72
Bouchard, Claude, 172
brain damage, 153–157
brain stem
    arousal system and, 179, 180f
    satiety and, 163–166, 164f
Brazil, 72–73
bread
    calorie density of, 84n, 231
    as one of top calorie sources, 83, 84
    satiety index and, 167–168, 169

breathing conditions, 185
Brobeck, John, 115
bromocriptine, 39–40
Brownell, Kelly, 54–55
Bucy, Paul, 199–200
buffet effect, 61–63
butter, 77–78

C
Cabanac, Michel, 138
cacao tree, 56
cafeteria diet, 19–20, 170, 223
caffeine
    as reinforcing, 55, 56n, 80–81
    reward system and, 232
    sleep and, 181
caloric return rates, 92–94
calorie density
    appetite management and, 231–232
    reinforcement and, 49
    sating ability of food and, 168, 168n
calories
    changing amount of physical activity
        and, 17–18
    changing intake of, 13–17, 14f, 17f
    history of, 11–13
    innate food preferences and, 50–53
    lack of balance between intake and
        output and, 18
    preferences and, 47–50
    top sources of, 83–84
Calories (kilocalories), defined, 12
calorimeters, 11–12
cannabinoid receptor type 1 (CB1), 64, 64n
canola oil, 77
carbohydrates
    conditioned flavor preferences and,
        48–49
    Dietary Guidelines for Americans and,
        1–2, 2n
    energy value of, 12n, 13
    lipostat and, 142–144
CB1 receptor. See Cannabinoid receptor
    type 1

cellular stress, 199n
cerebral cortex, 31, 31n
CFBAI. *See* Children's Food and
    Beverage Advertising Initiative
Chagnon, Napoleon, 72–73
Chee, Michael, 186
chemotaxis, 25n
Chestnut, Joey, 40
children, impulsivity and, 68n
Children's Food and Beverage Advertising
    Initiative (CFBAI), 227–228
chocolate, 56, 232–233
choice, economic
    manipulating, 225
    OFC neurons and, 103–105, 104f, 105f
    overview of, 217–218
cholecystokinin, 122n
Chow, Thomas, 180
cigarette smoking, 225
circadian desynchrony, 193–194
circadian rhythm disruption, 189–194, 233
c-Jun N-terminal kinases, 154n
coca plant, 55–56
Coca-Cola, 227
cocaine
    addiction and, 55–56
    dopamine and, 36–37
    freebasing and, 55n
    impulsivity and, 67
    locomotion and, 37f
cognition, 128, 182
Coleman, Doug, 121
Collins, Francis, 175
comfort food, 211–214
commodity crops, 226
computer analogy, 215–216
conditioned flavor preferences, 48, 48n,
    50n
control, lack of, 206–207, 206n
convenience, 226
cooking methods, 74, 81
corn, 94n
corn oil, 226
corn starch, 226

corn subsidies, 226
cortex. *See also* Orbitofrontal cortex
    cerebral, 31, 31n
    insular, 161
    prefrontal, 35f, 103, 104f
    sensory, 104n
    temporal lobe of, 200
    ventromedial prefrontal, 166n
corticosterone, 122n
corticotropin-releasing factor (CRF),
    203–205, 204f, 207, 207n
cortisol, 204–205, 204f, 208–211
cost-benefit decisions, 101–105, 102f
countermarketing, 224–225
crack cocaine, 55–56, 67
cravings
    advertising and, 85–86, 227
    controlling, 57
    loss of, 221
CRF. *See* Corticotropin-releasing factor
crossover design, 177n
cuckoo, common, 83, 83n
culture, 52, 94
Cushing, Harvey, 208
Cushing's disease, 208
Cushing's syndrome, 208, 209, 210

D
Dallman, Mary and Peter, 211–212
Dawkins, Richard, 42n
de Arujo, Ivan, 48–49
de Castro, John, 58
decerebrate rats, 163–164, 166
decision-making process
    in complex world, 24–26
    humans and, 29–32, 33–36, 35f
    lampreys and, 26–29
    learning and, 41–44
    sleep and, 186–188
decisiveness, 25–26, 27
delay discounting
    addiction and, 109–110n
    episodic future thinking and, 110–111
    overview of, 108–110

delayed rewards, 108–110

Denmark, 225

desks, treadmill, 173, 173n

desserts, 83–84, 232

desynchrony, circadian, 193–194

Detroit, Michigan, 226

dexamethasone, 212

diabetes mice, 118

Dietary Guidelines for Americans (USDA), 1–3, 83

digital media, 185

dimmer switches, 191

Dinges, David, 182

diphtheria toxin, 150n

discounting, delay. *See* Delay discounting

distance, snacking and, 99

dogs, reinforcement and, 46

L-dopa treatment, 38–39

dopamine

    addiction and, 55

    basal ganglia and, 36–40, 37f

    conditioned flavor preferences and, 48–49

    debate over role in reinforcement, 44n

    encephalitis lethargica and, 179n

    habit-forming drugs and, 80

    learning and, 44–47

    motivation and, 107

    threat response system and, 202

dorsomedial hypothalamus, 189n

drugs, 152. *See also Specific drugs*

E

economic choice system

    manipulating, 225

    OFC neurons and, 103–105, 104f, 105f

    overview of, 217–218

economics, changes in food production and, 99–101, 101f

education, role of, 224

efficiency in food-seeking, 91–94

effort cost, 98

eggs, birds' preferences for, 82–83

electrical engineering, 180

Eli Lilly, 148

emotion, food and, 128

encephalitis lethargica, 178, 179f, 179n

The End of Overeating (Kessler), 77

endoplasmic reticulum stress, 157n

endorphins, pleasure and, 47

energy, measurement of, 11–13

energy balance equation, 12

energy expenditure, 91–94

energy homeostasis. *See also* Lipostat

    hypothalamus and, 132–135, 133f

    raising awareness of, 153

    reproduction, blood glucose and, 156

Ensure, 137–138, 137n, 138n

environment

    changing, 223

    fixing, 230–231

    impact of, 175, 207

Epel, Elissa, 210

episodic future thinking, 110–111

Epstein, Leonard, 65–68, 109–110n, 110–111

Erdheim, Jakob, 115

ethical considerations, 152

evolutionary mismatches, 5, 70

exaptation, 31, 32

excitotoxicity, 80n

exercise

    benefits of, 142n, 234

    calories burned and, 17–18

    lipostat and, 140–142

extended amygdala, defined, 201

extreme obesity, 9, 113, 174, 229n

F

facial expressions, 202

familial fatal insomnia, 182–183

Family structure, 3n

Farooqi, Sadaf, 127, 131, 152, 173–174

fast food

    restricting access to, 226, 227

    spending on, 75, 76f, 100

fat (body)
  caloric content of, 14–15
  loss of vs. weight loss, 142
fats (dietary)
  conditioned flavor preferences and, 49
  Dietary Guidelines for Americans and,
    1–2, 2n
  energy value of, 12n, 13
  history of in modern diet, 77–78, 79f
  nutrition labeling and, 224, 224n
  sating ability of food and, 168–169
  sugar vs. as problem, 79
fatty acids, 124n
fearlessness, 200
feedback systems
  leptin and, 132–135, 133f
  negative, 132n, 211n
fever, 136
fiber, sating ability of food and, 169
fidgeting, 173
fight-or-flight response, 202–203
fish sauce, 80
flavor preferences, conditioned, 48, 48n,
    50n
flip-flop switches, 180
f.lux utility, 191
fMRI. *See* Functional magnetic resonance
    imaging
Foltin, Richard, 64
food assistance programs, 226
food production, 14
food reward. *See also* Reward system
  awareness of, 232–233
  importance of, 58–60
  influence on eating habits, 83–84
  in United States, 74–82
food scarcity, 52
food variety, 61–63
foraging, 87–94
Fox, Vicente, 226
fraternal twins, 171–172
free fatty acids, 122n
free will, 36n
freebasing, 55, 55n

French fries, 85
Friedman, Jeff, 125, 126n, 129–130, 152,
    158
Fröhlich, Alfred, 114
Fröhlich's syndrome, 114–115, 117
frontotemporal dementia, 106n
fruit, 167–168
full-spectrum light bulbs, 191
functional magnetic resonance imaging
    (fMRI)
  blood oxygen and, 160n
  brain responses to viewing images of
    foods and, 160–162, 162f
  sleep restriction and, 183, 186, 187

G
gamma-aminobutyric acid (GABA), 149n
garum sauce, 80
gastric bypass surgery, 221
Gearhardt, Ashley, 54–55
gene expression analysis, 153–154
General Mills, 227
genes, natural selection and, 42n
genetics, role of, 171–175, 218–219
glasses, blue-spectrum blocking, 191
globus pallidus, 30f
glucocorticoids, 209n
glucose, 48–49, 156
glutamate
  food preferences and, 49–50
  history of in modern diet, 78–80
gluttony, hunter-gatherers and, 94–97
glycerol, 122n, 124n
goals, learning and, 41–42
grain-based desserts, 83, 84
Grill, Harvey, 163–164, 166
Grillner, Sten, 23, 30
grocery stores, variety of food in, 75
growth hormone, 122n
Gurney, Kevin, 33

H
habituation, 62–63
Hadza people, 87–90, 88n, 90f, 91f, 94–97

Hall, Kevin, 14–16, 16n, 224
Hardee's, 228–229
Harris, Ruth B., 121–122
Heiman, Mark, 148
Helmchen, Lorens, 9
Henderson, Max, 9
Hervey, Romaine, 118–120, 120f, 133
Hetherington, Albert, 115, 148–149
high-fructose corn syrup, 76, 226
high-protein diets, 142–144
high-satiety foods, 167–171
Hill, Kim, 94–96, 94n
hippocampus, 200
Hirsch, Jules, 124, 129–130
holidays, 158, 159
Holt, Susanna, 167–170
homeostasis. *See also* Lipostat
    hypothalamus and, 132–135, 133f
    raising awareness of, 153
    reproduction, blood glucose and,
        156
honey, 95
hormones, 120, 208–211. *See also* Leptin;
    Specific hormones
HPA axis. *See* Hypothalamic-pituitary-
    adrenal axis
hunger, hunter-gatherers and, 95–96
The Hungry Gene (Shell), 126n
hunter-gatherers. *See also Specific groups*
    economic choice system and, 87–90
    optimal foraging theory and, 92–94
    overeating and, 94–97
    Paleolithic diet and, 170
    vegetables and, 92, 96n
    vitamin and minerals and, 52, 52n
hydrolyzed soy protein extracts, 78–79
hydrolyzed yeast protein extracts,
    78–79
hypothalamic-pituitary-adrenal (HPA)
    axis
    anatomy of, 204f
    comfort food and, 211–212
    cortisol and, 211n
    threat response system and, 203–204

hypothalamus. *See also* Ventromedial
    hypothalamic nuclei
    anatomy of, 113f
    arcuate nucleus of, 146–148, 147f
    arousal system and, 179, 180f
    changes in with obesity, 153–158
    cortisol and, 209
    dorsomedial, 189n
    economic choice system and, 104n
    food intake regulation and,
        166–167
    functional magnetic resonance imaging
        and, 162n
    homeostasis and, 132–135, 133f
    lateral, 168, 180f
    lipostat and, 218
    long-term energy balance and adiposity
        and, 166
    MSG and, 80n
    neuropeptide Y and, 146, 148
    palatability, satiety index and, 168
    pituitary hormones and, 114n
    tumors on, 113

I
identical twins, 171–172
Ikeda, Kikunae, 80
impulse control, 109n
impulse control disorder, 38–39
impulsivity, 67, 68n, 109n
industrialization
    food processing, distribution, and
        preparation and, 69–70
    obesity and chronic disease and, 8–11,
        10f
    Yutala and, 7–8
infant formula, 228
inflammation, 153–154, 154n, 157
information processing, 215–216
innate food preferences and aversions,
    51
input cues, impacts of on outputs, 216
insomnia, familial fatal, 182–183
insular cortex, 161

insulin
  energy balance and, 146n
  neuropeptide Y and, 146–147
  satiety system and, 122n
  ventromedial hypothalamic nucleus
    and, 117n
insulin resistance, 153–154
INTERSALT study, 73
intestines, 164–165
intralaminar nucleus, 166n

J
Jim (former miner), 39–40

K
Kahneman, Daniel, 3–4, 36, 216
Kalahari desert, 70–72
Kalra, Satya, 146
Kennedy, Gordon, 120
Kessler, David, 77
ketogenic low-carbohydrate diets, 143n
ketone bodies, 122n
Keys, Ancel, 123, 128–129, 128n
kilocalories, 12
King, Bruce, 117n
Kitava, New Guinea, 7–8
Klüver, Heinrich, 199–200
Klüver-Bucy syndrome, 200
Kraft, 227
Kratz, Mario, 2n
!Kung San people, 70–72

L
labeling, 224, 224n
lampreys, 23–29, 23n, 24f, 25n
Landen, Jeremy, 76
lard, 78
lateral hypothalamus (LH), 168, 180f
laterodorsal tegmentum (LDT), 180f,
    181f
laziness, 107
LC. See Locus coeruleus
L-dopa treatment, 38–39
LDT. See Laterodorsal tegmentum

lean mass, metabolic rate and, 15n
learning
  decision-making process and, 41–44
  dopamine and, 44–47
  threat response system and, 205
Lee, Richard, 70–71
legislation, 229
Leibel, Rudy, 122–126, 129–130, 131, 152,
    159, 183
leptin
  corticosteroids and, 209n
  discovery of, 125–126
  feedback and control of, 132–135, 133f
  function of, 126–130
  hypothalamus and, 167
  inflammation and, 154
  leptin resistance and, 158
  long-term energy balance and adiposity
    and, 166
  neuropeptide Y, melanocortins, and
    regulation of, 146–147
leptin deficiency, 127–129
leptin resistance, 134, 158, 167, 209
leptin signaling pathway mutations,
    173–174
leptin therapy, 131, 131n
Levin, Barry, 136–138, 140
Levine, James, 173
LH. See Lateral hypothalamus
light bulbs, 191–192, 233
Lindeberg, Staffan, 7–8
lipodystrophy, 131n
lipostat
  appetite management and, 231–232
  carbohydrates and, 142–144
  changing set point of, 135–140
  data supporting existence of, 122
  exercise and, 140–142
  Hervey's model of, 120, 120f
  leptin deficiency and, 128
  melanocortins and, 148
  neuropeptide Y and, 146–147
  overview of, 132–135, 134f, 218
  regulation of, 157–158

lipostat (*continued*)
sleep restriction and, 183
threat response system and, 209–210
location of dining, at home vs. out, 75, 76f
locus coeruleus (LC), 180f
low-carbohydrate diets, 142–144, 143n, 146n
Lowell, Brad, 151, 166

M
machine feeding study, 58–59, 59f
Maduru and Esta (Hadza people), 87–90
magnetic resonance imaging (MRI), 156, 160
marijuana, 63–65
Marlowe, Frank, 90
marshmallow experiment, 108–109, 109n
McDevitt, Ross, 44–46
McDonalds, 228
meal size, 166
meat, 167–168
mediation, 234–235
melanocortin-4 receptor, 174
melanocortins, 148–149, 148n
α-melanocyte-stimulating hormone, 148n
melatonin, 189–190, 191
Melhorn, Susan, 160
Menaker, Michael, 192
Mendonca, Suzanne, 135
mescaline, 200
metabolic rate, increasing, 13n
metabolic wards, 20–21
methylprednisolone, 208–209
Mexico, 225–226
microarrays, 153–154
microglia, changes in with obesity, 154–155, 155f
milk, 2, 2n, 228
mindfulness meditation, 234–235
minerals, lack of enjoyment of, 52, 52n
Minnesota Starvation Experiment, 123, 128–129
Minnie G., 208
mismatches, evolutionary, 5, 70

Mohr, Bernard, 113–114, 142, 152
mongongo trees, 70–71, 71f
monosodium glutamate (MSG), 49–50, 78–80, 80n
Moser, Elisa, 113, 117
Moss, Michael, 77, 86n, 100
motivation
bland diet and, 139
brain regions and, 40
dopamine and, 107
learning and, 42–44
MRI. *See* Magnetic resonance imaging
MSG. *See* Monosodium glutamate

N
natural flavors, 80
natural selection, 42n, 96
Nature papers, 125, 126n
NEAT. *See* Non-exercise activity thermogenesis
negative feedback systems, 132n, 211n
negative reinforcement, 44
Nestlé, 228
Nestle, Marion, 84n
Neumann, Rudolf, 123
neuron-firing, valuation and, 27n
neurons, subjective value and, 103–105
neuropeptide Y neurons (NPY neurons), 149–151, 149n, 151f, 166
neuropeptide Y (NPY), 146–149, 209
New Guinea, 7–8
NGY/AgRP neurons, 149n
night blindness, 61n
night eating syndrome, 194
noncontact aggression, 206
non-exercise activity thermogenesis (NEAT), 173
nonindustrial diets. *See also* Hunter-gatherers
common elements of, 73–74
fat, starch, sugar, salt, free glutamate and, 76
noradrenaline, 202
Norgren, Ralph, 163

nose-poking experiment, 44–46
NPY. *See* Neuropeptide Y
NPY neurons. *See* Neuropeptide Y
    neurons
NTS. *See* Nucleus tractus solitarius
NTS neurons, 166
nucleotides, umami flavor and, 49n
nucleus accumbens, 33n
nucleus tractus solitarius (NTS), 165, 165f,
    166
nutrition labeling, 224, 224n
nuts, 232

O
oatmeal, 168
ob gene, 118, 121, 124–125, 146–147
obese mice, 118, 119f, 121–122, 123, 150
obesity, statistics on increase in, 9–11, 10f
obesity epidemic, defined, 11
OFC. *See* Orbitofrontal cortex
OFT. *See* Optimal foraging theory
oils, 78
opportunistic voracity, 21
optimal foraging theory (OFT), 92–94,
    92n
optimism bias, 186, 219
optogenetics, 149n
O'Rahilly, Stephen, 126–127, 131, 152,
    173–174
orbitofrontal cortex (OFC)
    anatomy of, 104f, 105f
    economic choice system and, 103–105
    functional magnetic resonance images
        of, 161, 162f
    nonfunctional, 105–106
    parabrachial nucleus and, 166n
overall food motivation, 65n
overfeeding experiments, 118–122,
    131–132

P
Padoa-Schioppa, Camillo, 101–105
palatability, 57–63, 138, 168
Paleolithic diet, 170

pallium, lamprey, 27–28, 31, 31n, 32f
Palmiter, Richard, 37f, 107, 150–151,
    166
Papua New Guinea, 7–8
parabiosis studies, 118–122, 119n
parabrachial nucleus, 166, 166n
parasitism, 83n
paraventricular hypothalamic nucleus,
    203n
Pardi, Dan, 186
Parkinson's disease, 36–38
patent application, leptin and, 125–126,
    126n
Pavlov, Ivan, 46, 57
Pavlovian conditioning, 46
pedunculopontine tegmentum (PPT),
    180f, 181f
perifornical area (PfF), 181f
perseveration, 106, 106n
physical activity, 17–18, 173, 234. *See also*
    Exercise
physiology, brain as computer and, 216
pineal gland, 189
pituitary gland, 113, 113f, 114–115, 208
pituitary hormones, 114n
planning, importance of, 57
plateaus, weight-loss, explanation for, 16
pleasure, dopamine, endorphins and, 47
plovers, ringed, 82
polycose, 48n
polysomnography, 177n
POMC neurons, 148–151, 151f, 166
Pontzer, Herman, 91, 94–95, 95n
Porte, Dan, 145
positional cloning, 125n
potato diet, 60–61, 61n, 171
potatoes, 60–61, 61n, 167–168, 171, 231
PPT. *See* Pedunculopontine tegmentum
prednisone, 208
prefrontal cortex
    anatomy of, 104f
    decision-making process and, 35f
    economic choice system and, 103
    parabrachial nucleus and, 166n

protein
    conditioned flavor preferences and, 49
    energy value of, 12n, 13
    lipostat and, 142–144
    sating ability of food and, 169
psychic akinesia, 39–40, 39n
psychological stress
    brain and, 199–205
    comfort food and, 211–214
    food intake and, 206–210
    managing, 234–235
    overview of, 197–198
public health measures, 224
Puzder, Andrew, 228–229

R
Ranson, Stephen, 115, 148–149
Ravussin, Eric, 20–21, 98, 208–209, 223
recessive inheritance, 118n
Redgrave, Peter, 33, 36n
regulation of food industry, 227–229
reinforcement
    addiction and, 53–57
    caffeine, alcohol and, 80–81
    caloric density and, 49
    debate over role of dopamine in, 44n
    dopamine and, 46
    nature of food causing, 47–50
    overview of, 43–44
    valuing, 66–67
reinforcement pathology, 67
relative reinforcing value of food
        (RRV$_{food}$), 66–67
reproduction, 96, 156
response inhibition, 109n
restraint stress, 212
reverse-marijuana, 64–65, 64n, 152
reward system. See also Food reward
    overview of, 217
    restricting, 139
    stress response and, 206–207, 213
Rift Valley, 87–90
rimonabant, 64–65, 64n, 152
ringed plovers, 82

risk, 94, 186–188
RNA microarrays, 153–154
Rolls, Barbara, 62
rotating shift work, 192–193
Roux-en-Y gastric bypass, 221
RRV$_{food}$. See Relative reinforcing value of
        food
Rudd Center for Food Policy and Obesity,
        227, 228

S
Salamone, John, 107
salt, 51n, 52n
Salt Sugar Fat (Moss), 77, 86n, 100
Saper, Cliff, 179–180, 182
satiation, defined, 164n
satiety, 62–63, 164n
satiety center, 117, 163–166. See also
        Ventromedial hypothalamic nuclei
satiety factor, 118–126. See also Leptin
satiety index, 167–171
satiety system, 118–126, 218, 231–232
saturated fat, 157
schools, 226
Schur, Ellen, 156, 160–163, 183
Schwartz, Mike, 5, 145–149, 151–153, 154
Science papers, 147–148
Sclafani, Anthony, 19, 47–50, 223
SCN. See Suprachiasmatic nucleus
SCT-Orange safety glasses, 191
Seeley, Randy, 148, 221
selection problem. See also Decision-
        making process
    basal ganglia and, 26–29, 28f
    defined, 24
    human solution to, 29–32, 29f, 30f
    lampreys and, 25, 25n, 26–29
    overview of, 24–26
selectors, properties of, 25–26
The Selfish Gene (Dawkins), 42n
self-regulation of advertising, 227–228
self-reporting, problems with, 20
semistarvation neurosis, 128
sensory cortex, 104n

sensory-specific satiety, 62–63

serotonin, 202

settling point, set point vs., 134n

Shell, Ellen Ruppel, 126n

shift work, 192–193

Siffre, Michel, 188

Sims, Ethan, 124, 124n, 131–132

sleep
  biological basis of, 178–183, 179f, 180f,
      181f
  circadian rhythm disruption and,
      189–192
  differing requirements for, 185
  impacts of restricting, 177–178, 182–184
  importance of, 219
  long periods of, 184n
  prioritizing, 233
  risk and reward perception and,
      186–188

sleep apnea, 185, 233

sleep center, 179

sleep disorders, 185

snacking, distance and, 99

SNAP. *See* Supplemental Nutrition
    Assistance Program

social hierarchies, 206

SOCS3 protein, 154n

sodas, 56n, 76, 225–226

sodium, 51n

soy sauce, 80

soybean oil, 77, 78, 226

soybean subsidies, 226

Space Race, 188

Stanford marshmallow experiment,
    108–109, 109n

starches, 61n

starvation, comparison to leptin
    deficiency, 128–130

starvation response, 129–130, 221

STAT3, 209n

Stephenson-Jones, Marcus, 27n, 30

stereotaxic apparatus, 115, 116f

Sternson, Scott, 149, 151

stomach, 164–165, 168

St-Onge, Marie-Pierre, 177–178, 183

stress, 199, 199n. *See also* Psychological
    stress

Stress in America Report (2007), 198

striatum
  bid strength and, 26n
  decision-making process and, 35f
  human, 30f
  lamprey, 26–27, 27n
  orbitofrontal cortex and, 104n
  ventral, 33, 38–39, 45f, 46, 160–161, 162f

stroke, 106n

subjective value, 102–103

subparaventricular zone, 189n

subsidies, 226

substantia nigra, 30f, 36–38

subthalamic nucleus, 30f

sugar
  fat vs. as problem, 79
  history of in modern diet, 76–77, 77n,
      78f

supermarket diet, 19. *See also* Cafeteria
    diet

supernormal stimuli, 82–83

supersizing, 228

Supplemental Nutrition Assistance
    Program (SNAP), 226–227

suprachiasmatic nucleus (SCN), 189–190,
    190f, 193, 194n

surgery, weight-loss, 221

sweeteners, 76, 78f. *See also* Sugar

sympathetic nervous system, 203f

system 1 brain processes, 3–4, 216–217

system 2 brain processes, 4, 216

T

Tanzania, 87–90

taste information, 161, 213

taxation, 225–226

technology, 191–192

temperature regulation, 132–134, 133f, 136

temporal lobe of cortex, 200

tetrahydrocannabinol (THC), 64

thalamus, 31n, 104n, 166n, 180f

Thaler, Josh, 154
THC. *See* Tetrahydrocannabinol
theobromine, 56, 232–233
thermogenesis, non-exercise activity, 173
thermostat analogy, 132–134, 133f
Thinking, Fast and Slow (Kahneman), 3–4, 216
Thorndike, Edward, 43
threat response system, 199–205, 202f, 212–213, 219–220
thyroid hormones, 122n
timing of meals, 193
Tinbergen, Nikolaas, 82
TMN. *See* Tuberomammilary nucleus
tobacco industry, 225
touch, 215n
traditional diets, 170
trans fat, 224, 224n
treadmill desks, 173, 173n
tuberomammilary nucleus (TMN), 180f, 181f
Turek, Fred, 192, 193
Twilight app, 191
twin studies, 171–172

U
Ulrich-Lai, Yvonne, 213, 214
umami flavor, 49–50, 49n, 78
uncontrollable stressors, 206–207, 206n
urocortins, 203n
U-Select-It 2007 vending machines, 20–21, 98

V
vagus nerve, 165, 165f
valuation
  calculation of, 101–105, 102f
  of future self, 108–110
  neuron-firing and, 27n
variety, food, 61–63
vegetable oils, 78
vegetables, hunter-gatherers and, 92, 96n
Velloso, Licio, 153–154
vending machine studies, 20–21, 98

Venezuela, 72–73
ventral periaqueductal gray (vPAG), 180f, 181f
ventral striatum, 33, 38–39, 45f, 46, 160–161, 162f
ventral tegmental area (VTA), 30f, 38, 44, 45f, 160–161, 162f
ventrolateral preoptic area (VLPO), 178n, 179, 181, 181f
ventromedial hypothalamic nuclei (VMN), 115–117, 115n, 116f, 117f, 117n, 119–122, 148–149
ventromedial prefrontal cortex, 166n
Vincent, Ali, 135
vitamin A deficiency, 61n
vitamins, lack of enjoyment of, 52, 52n
VLPO. *See* Ventrolateral preoptic area
VMN. *See* Ventromedial hypothalamic nuclei
Voigt, Chris, 60–61
von Economo, Constantin, 178–179, 179f
voracity, opportunistic, 21
vPAG. *See* Ventral periaqueductal gray
VTA. *See* Ventral tegmental area

W
Wansink, Brian, 98
weight loss, fat loss vs., 142
weight-loss plateau, explanation for, 16
weight-loss surgery, 221
Weigle, Scott, 143
Westerterp, Klaas, 143
Westerterp-Plantenga, Margriet, 143
wheat subsidies, 226
white bread, 167–168, 169
white flour, 226
whole-grain bread, 167–168, 169
Why Calories Count (Nestle), 84n
willpower, 172n
Wilson, Mark, 205–207
Winterhalder, Bruce, 93
Wise, Roy, 42, 53
withdrawal, 53n. *See also* Cravings
Wood, Brian, 90, 94–95

Woods, Steve, 145
World War II, 123

Y
Yanomamö people, 72–73
Yutala (Kitavan), 7–8, 220

Z
Zakrzewska, Katerina, 209
Zeitzer, Jamie, 186
zoning laws, 226
Zucker, Lois and Theodore, 118
Zucker fatty rats, 118, 122